T0183975

Dynamic Secrets in Communication Security

Sheng Xiao · Weibo Gong
Don Towsley

Dynamic Secrets in Communication Security

 Springer

Sheng Xiao
College of Information Science
  and Engineering
Hunan University
Changsha, Hunan
People's Republic of China

Don Towsley
Department of Computer Science
University of Massachusetts Amherst
Amherst, MA
USA

Weibo Gong
Department of Electrical and Computer
  Engineering
University of Massachusetts Amherst
Amherst, MA
USA

ISBN 978-1-4939-0229-3   ISBN 978-1-4614-7831-7   (eBook)
DOI 10.1007/978-1-4614-7831-7
Springer New York Heidelberg Dordrecht London

Printed on acid-free paper

Springer is part of Springer Science+Business Media (www.springer.com)

Preface

This book introduces *dynamic secrets* as a new approach for cryptographic key management in secure communication systems. This approach allows a stolen cryptographic key to be automatically recovered shortly after the incidence of key theft. It also provides an intrinsic mechanism to promptly detect *all* active attacks using a stolen key.

Dynamic secrets complement and expand Kerckhoffs' principle, a guideline for cryptosystem design since the nineteenth century. Kerckhoffs' principle asserts that security of a cryptosystem should solely rely on the secrecy of its key. In this monograph, we demonstrate that dynamic secrets can constantly "refill" key secrecy during the communication process. Therefore, a secure communication system can remain functional even if its key is known to an adversary.

The intended audiences of this book include researchers, professionals, graduate and undergraduate students with interests in communication security. In order to rigorously analyze security properties provided by dynamic secrets, this book contains ideas from information theory and probability theory. We put efforts in writing to ensure that readers who are unfamiliar with these disciplines can perceive the entire scope of dynamic secrets, and implement dynamic secrets in practical secure communication systems after reading this book.

Amherst, MA, USA, January 2013

Sheng Xiao
Weibo Gong
Don Towsley

Acknowledgments

The author Sheng Xiao thanks the other two authors, Prof. Weibo Gong and Prof. Don Towsley for their mentorship and support throughout his Ph.D. years. Without them, the dynamic secrets research would be impossible. He would also like to thank Prof. Dennis Goeckel and Prof. Hossein Pishro-Nik in University of Massachusetts, Amherst, Prof. Robert Gao in University of Connecticut, Storrs, Prof. Jie Wang in University of Massachusetts, Lowell, and Prof. Xiaohong Guan in Xian Jiaotong University, China, for their helpful discussions and encouragements for the research. He also wishes to thank Dr. Roberto Padovani and Prof. Jack Wolf for their encouragement when they visited UMASS.

The author Weibo Gong thanks Prof. Michael Rabin whose work on everlasting security motivated the initial thought on using channel randomness for security. He would also like to thank Prof. David M. Pozar for the early discussions on channel randomization for secure wireless communications.

All three authors thank the United States National Science Foundation for their support of grants EFRI-0735974, CNS-1065133, CNS-1239102, and CNS-1018464, the Army Research Office for their support of contract W911NF-08-1-0233, and the University of Massachusetts for their support of the CVIP Technology Award Fund.

Contents

Chapter 1
Introduction

There are two methods for identifying a person. If that person is a complete stranger, we have to rely on his identity document (ID), such as a driver license, to identify him. If he is our acquaintance, we do not check his ID. Instead, we identify him by familiarity. Familiarity includes our perception of his body figure, his face, his voice, and other biometrics that allow us to recognize him.

More importantly, familiarity includes the memories shared between that person and us. Even if we cannot see him or hear his voice, we still can easily identify the person by asking him about the historical information that both him and we would remember.

For example, during wedding ceremonies, there is a popular game called "pick your man". The bride has her eyes folded. The groom and some volunteer guests stand in a row. The bride asks yes-no questions. Every participant answers the question by nodding or shaking head. The guest who has an incorrect answer leaves the game after each round. Quickly the bride can eliminate the guests and leave the groom outstanding. Exceptions do occur when the groom forgets information that the bride believe to be very important and that he should remember. This is the fun part of the wedding but not the point that we want to emphasize in this monograph.

Suppose the bride and the groom both have good memories, their shared information from the past will be solid basis for the identification. The more familiar two people are, the more reliably they can recognize each other. If two people have shared a large amount of information in the past, it becomes much more difficult to impersonate one in front of the other, i.e. defeat the identification by shared history, than to fake a driver license, i.e. defeat the identification by ID.

We do not have similar experience in secure communications in the digital world, such as the authentication to login our email accounts. We may login to our email accounts multiple times a day from our laptop computers, repeat the logins everyday in several years. The email server asks for our usernames and passwords every time. The email server and our laptops refuse to become acquaintances although they have exchanged many gigabytes of data, much more than a human can exchange with another in day to day life. This large amount of information should have formed a stronger proof of our identities than usernames and passwords. Had we have daily conversation with a person for the past several years, we would not expect him to

S. Xiao et al., *Dynamic Secrets in Communication Security*,
DOI: 10.1007/978-1-4614-7831-7_1,
© Springer Science+Business Media New York 2014

check our IDs every time we talk. He should be able to recognize us more reliably by our communication history.

There is a mismatch between the typical digital authentication process and the possibly better alternative approach of identification between acquaintances. *This monograph is about methods to establish and make use of "familiarity" between digital communication devices to improve communication security.*

Specifically, we propose to generate *dynamic secrets* from the communication history between a pair of communication devices. Dynamic secrets are a sequence of hash values. For every small period of time, each of the two devices generates a dynamic secret by hashing all the information communicated between them during this period. Both devices generate the same dynamic secret because it is the hash value of their shared information in that period of time.

During the communication process throughout a long period of time, many dynamic secrets are generated. The two devices do not keep all the dynamic secrets. Instead, they keep the XOR value of all the dynamic secrets have been generated. The XOR value is a highly compressed digest of the communication history of the two devices. Effectively, the XOR value of the dynamic secrets represents the "familiarity" between two devices. We denote the XOR value of the dynamic secrets generated up to time t as $S(t)$.

Suppose these two devices use symmetric key encryption to protect the data confidentiality in their communications. They have a pre-shared symmetric key $k(0)$. In traditional secure communication schemes, $k(0)$ and session keys derived from $k(0)$ will be used in the data encryption. In contrast, dynamic secrets will turn the traditional key $k(0)$ into a dynamic key $k(t)$ whereby $k(t) = k(0) \oplus S(t)$. $\oplus$ is the bitwise-XOR. Here we assume $S(t)$ has the same binary length as $k(0)$.

Whenever a dynamic secret is generated, the XOR value of dynamic secrets, $S(t)$ will be updated. At the same time, the two devices update the symmetric key. Therefore the key is constantly changing during the communication process.

The dynamic key $k(t)$ is resistant to key thefts. Even if the adversary steals the initial key $k(0)$, he can only use $k(0)$ to decrypt the communications occurred in a small period of time starting from time 0. The value of $k(t)$ is constantly changing and determined by both $k(0)$ and all the communications between the two devices from time 0 to time t. The adversary is then forced to keep eavesdropping *all* communications from time 0 to time t to keep tracking the updates of $k(t)$. Such an eavesdropping process is not only costly, but also substantially increase the chances of exposing the adversary. Moreover, in many practical communication scenarios, such as mobile wireless communications, error-free eavesdropping for a long period of time is practically impossible. The adversary would have much less incentive to steal a dynamic key than to steal a traditional key.

$k(t)$ is also resistant to duplication. Suppose the adversary duplicates the dynamic key used between two devices. The adversary then impersonates as device A to communicate with device B. This fraudulent communication will produce a change in the key of device B. The key of device A remains unchanged. Later when devices A and B resume their communications, the key disparity will surely uncover the existence of the impersonating adversary.

Now, we may hypothesize an email login system using dynamic secrets and dynamic keys. As my laptop computer powers on, the email client software on my laptop computer connects to the email server and requests access to the emails stored in my account. After a mutual authentication using the dynamic key, my laptop computer begins to fetch new emails. The communications between my laptop computer and the email server are encrypted by the dynamic key and constantly change the dynamic key. The entire process is automatic and secure.

The password theft adversary is a much less threat than in the traditional username-password authentication scheme. Even if the adversary clones my laptop computer and temporarily possesses the dynamic key. He cannot hide and quietly read my emails. Any of his connection attempt to the email server will expose his existence. Moreover, my next login to the email server will expire the previous dynamic key and destroy his malicious efforts of cloning my laptop computer. As long as I can keep my laptop under control, the access to my email account is guaranteed to be protected.

The change in email systems is just a small example. Dynamic secrets can bring many interesting properties to various existing secure communication systems. The analysis to dynamic secrets has the potential to open a series of topics in the communication security research. The rest of this monograph is organized as follows:

In Chap. 2, we demonstrates a common bottleneck of many existing secure communication systems. The communication security of these systems relies on the safety of the cryptographic keys, which is hard to protect in practice. We summarize the paradigm of these existing secure communication systems as "security by ID", as opposed to our idea of "security by communication history". Chapter 3 defines the dynamic secrets generation and the dynamic key update processes in a noisy communication model. This chapter also explores three unique security features brought by dynamic secrets: true randomness at low cost, automatic stolen key recovery, and inherent detection to impersonation attacks. Chapter 4 implements dynamic secrets in an office wireless LAN test platform. The implementation details and performance issues are discussed. Chapter 5 explores the characteristics of the key management system based on dynamic secrets. The discussions in this chapter take the smart grid as the application scenarios and compares the dynamic key management with traditional key management. Chapter 6 introduces a theoretic foundation of dynamic secrets and distinguishes the dynamic keys from traditional, relatively static cryptographic keys. This chapter also discusses the connections between dynamic secrets and various research domains such the random hashing, quantum key distribution, one time pad encryption, the channel coding in communications. Chapter 7 presents a reliability analysis view of communication security. Interestingly, the reliability engineering theory which is traditionally used to measure the performance consistency of mechanical systems can be used to evaluate the performance of secure communication systems in the presence of key theft problem. The analysis shows that dynamic secrets can complement and improve traditional secure communication systems by enhancing several reliability measures that were previously ignored. In Chap. 8, we propose several potential applications of dynamic secrets.

Chapter 2
Communication Security and Key Safety

In order to allow a secure communication system to function properly, users must keep some secret, which is often referred as a cryptographic key or in short, a key. Safety of key is a premise of communication security because once adversary knows the key, the users' communications will no longer be protected.

For centuries, the above understanding of the relationship between key safety and communication security holds as an axiom. However, the prerequisite of key safety is difficult to guarantee in practice. There are numerous vulnerabilities that allow the adversary to obtain the key and then compromise communication security. The decisive role of key safety becomes an inherent weakness in many practical systems.

This entire monograph is devoted to present dynamic secrets as an approach to relieve the tension between the possibility of key theft and the demand for communication security guarantee. In this chapter, we briefly review typical secure communication systems and their design principles. This background review explains the motivation and the basis of our research.

Section 2.1 presents a series of historical ciphers and then reviews the Kerckhoffs' guidelines for cryptosystem design. These guidelines have served as a main frame for many secure communication systems from nineteenth century to modern days. We compare Kerckhoffs' guidelines with the art of locksmith. The comparison demonstrates that the key safety is a single point of failure to communication security. Section 2.2 illustrates the practical challenges of key safety protection by presenting a collection of attacks that allow adversary to obtain key.

2.1 Secure Communication System Design and Locksmith

The history of secure communication systems can be dated back at least 2,500 years ago. The Spartans invented a cipher namely *scytale* and used this cipher to protect confidential messages transfer in war time. As shown in Fig. 2.1a, a scytale consists of a wooden rod and a strip of parchment. The message sender wraps the parchment

S. Xiao et al., *Dynamic Secrets in Communication Security*,
DOI: 10.1007/978-1-4614-7831-7_2,
© Springer Science+Business Media New York 2014

(a)   **(b)**

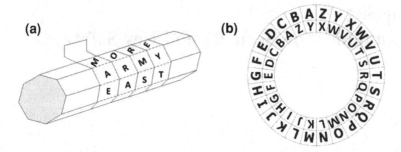

Fig. 2.1 Ancient systems for secure communication: **a** a scytale, **b** a Caesar's cipher disk

strip around the rod and then writes his message along the rod's axis direction. The unwrapped parchment strip is carried by a messenger to the recipient, who has a rod of the same diameter as the rod used by the sender. The recipient re-wraps the parchment strip and read the message. The rod diameter is a secret between the sender and the recipient and blocks their enemy from read the message.

The famous Julius Caesar used another method to secure the communications in his army. When writing a confidential message to his generals, Caesar substitutes every letter by a different letter in the alphabet. The substitution rule was kept as a secret between Caesar and his generals. In this way, even if the encrypted message is intercepted by his enemies, they may discard the encrypted message as meaningless scrambled letters. The cipher that substitutes each letter with a letter some fixed number of positions down the alphabet is named after Julius Caesar as Caesar's cipher. Figure 2.1b shows a Caesar's cipher disk that helps quickly encrypt and decrypt messages.

There are many other ancient systems that provide communication security. Generally, these systems are fundamentally weak when judged by modern standard because they are easily crackable when adversary learns the system construction. For example, the scytale system is easily defeated if the enemy of Spartans knows that decryption is to re-wrap the parchment on a rod. It is not difficult for the enemy to make a rough guess of the rod's diameter and decrypt the messages. Caesar's enemy only needs to try several times to find the exact offset between the plaintext message alphabet and the encrypted message alphabet. Once the enemy finds the offset, the secure communication is no longer secure.

In 1883, Dutch linguist and cryptographer Auguste Kerckhoffs proposed six cryptosystem design guidelines [58]. He suggested to divide a cryptosystem into two parts: the key and the system. More importantly, he emphasized that the communication security must remain intact even if the system construction is known by enemy. In other words, Key safety is the sole premise of communication security. This design principle is later recognized as *Kerckhoffs' principle*.

Kerckhoffs' principle distinguishes modern cipher designs from ancient ciphers such as the scytale and Caesar's cipher. A famous modern cipher example is Enigma, used to protect the German military communications in World War II.

The Berlin headquarter encrypts messages using the Enigma machine and broadcast the encrypted messages through radio signals. Then the German submarine fleets, often referred as the U-boat, would perform their battle tasks according to the received messages.

Enigma is a complicated electric-mechanical device that consists of some wired boards and rotors. The key of encryption and decryption is the initial rotor positions. Berlin sender set rotors to some secret initial positions and feed the plaintext message into the machine. The output of Enigma machine is the encrypted message. The U-boat receiver will set his Enigma machine to some corresponding rotor initial positions and then input the encrypted message. The machine output will be the decrypted, plaintext message.

Capturing the Enigma machine alone does not allow the Allies to completely crack the German military secure communication system. It took some of the world's smartest minds such as Alan Turing working days and nights to analyze the structure and the cryptographic properties of the Enigma machine and decrypt some Enigma-encrypted messages. In order to efficiently decrypt the targeted German military communications, it was also vital to have special military missions that brought back the German codebooks that contain the initial rotor positions.

The cracking of the Enigma cipher was a grand victory. The Allies were then able to stop the German U-boats from sinking their Atlantic transport ships. The Allies were also able to send fake battlefield information to the German commanders and lure them to make wrong decisions. It was conjectured that the World War II would have been ended in 1948 instead of 1945, had the Enigma cipher not been cracked [117].

On the other hand, the cracking process of the Enigma machine is a proof of the effectiveness of Kerckhoffs' guidelines. The key-system separation design significantly increased the difficulty of cracking the Enigma cipher. It is necessary to have knowledge on both the system structure and the key to break the communication security. Many modern day secure communication system designs adopt Kerckhoffs' principle. The cryptographic algorithms and protocols used in the system are documented publicly and left key safety as the necessary and sufficient condition for communication security. Such a system is referred as an open cryptosystem in the context of communication security. Most civilian secure communication systems and a large number of military secure communication systems are open cryptosystems.

The open system design of secure communication systems reminds us of another craft with long history, the locksmith. Just like the key-system separation in the open system design of secure communication systems, a door lock consists of two components, the lock body installed on the door and the key. The search for strong cryptographic algorithms and protocols is like the search for lock structures that are resistant to lock-picking. With a strong lock body, key safety is the prerequisite for security. Key theft is disastrous to both secure communication systems and door locks.

Unlike lock body, which is often vulnerable to lock picking and brute force break in, modern day cryptographic algorithms and protocols can be extremely sophisticated and resistant to cryptanalysis attacks. A good example is AES Rijndael algorithm, the current NIST standard for electronic data encryption [91]. It was proposed in 1998 and standardized in 2001. No publicly known efficient cryptanalysis attack to AES Rijndael has been developed for more than a decade. The adversary who defeats AES by cryptanalysis must surpass all the public research efforts on AES throughout these years. The open system principle eliminates weak cryptographic designs by opening them up to the public tests.

However, it is more complicated to protect a cryptographic key than to safeguard a lock key. A lock key is a physical entity that the key owner can effectively check if it is stolen. Moreover, a lock key can be made with special three dimension structure and contains rare materials. Duplicating a carefully designed lock key is difficult and costly. Even if a thief knows the shape and build of the lock key, he may not be able to create one. On the other side, a cryptographic key is merely a piece of information. Unless adversary confessed or caused noticeable security damage, the cryptographic key owner would not be able to recognize that his key is known to adversary. The duplication of a cryptographic key only requires copying a bit string. An adversary may duplicate a cryptographic key remotely with negligible duplication cost.

In modern day secure communications, although cryptanalysis attacks are still a viable techniques to compromise the communication security, it is often more cost-effective for adversary to focus on obtaining the cryptographic key.

2.2 Challenges to Ensure Key Safety

Attacks that threaten key safety can be generally classified into two types. One type of attacks is key cracking. Adversary attempts to deduce the key from information available to him. For example, the adversary may analyze eavesdropped cipher texts that are encrypted from some known plain texts and try to calculate the key. Another type of attacks is key stealing. Adversary obtains the key through an unauthorized access to the key. In later chapters, we use *key theft* to represent the incident that adversary obtains the key, regardless of the type of attacks.

2.2.1. Key Cracking

Exhaustive search is a trivial yet effective key cracking attack. Adversary first eavesdrop a short segment of key related information such as the key's hash value in an authentication process, then he try out possible key values in the key space to find the value that produces the hash value. Exhaustive search attack is extremely effective against human memorable passwords. Research work in [126] estimates that the majority of human created passwords have less than 20 bits of entropy by NIST standard tests (NIST SP800-63). With today's computing technology, exhaustive search attack can reveal a large amount of passwords in several hours [68, 69].

The exhaustive search attack can be defended by generating the cryptographic key using a pseudo random number generator (PRNG) and then storing the key in a secure storage device. The pseudo random key values spreads in a large key space. The exhaustive search would take an unreasonably long time, e.g. more than a thousand years, to find the key.

PRNG is an algorithm that expands a short numeric seed to a long sequence of apparently random numeric values. The algorithm design defects, the implementation flaws, and the insufficient randomness in the seed value are vulnerabilities that an adversary may exploit for key cracking. For example, research works in [35] and [87] study the weakness in the key generation algorithms to predict the pseudo random key values with high probability. Reference [124] shows that various implementation flaws can shrink the key space considerably and allow the adversary to exhaustive search the key in a limited key space. A famous incident was the implementation flaw found in the OpenSSL library in Debian Linux operating system. A function that is supposedly to keep supplying entropy to the numeric of the PRNG has been neglected in the implementation. Therefore, the adversary may predict the outcomes of the PRNG and then explicitly calculate the cryptographic keys generated in the system [18].

The countermeasure to the exhaustive search attack and the PRNG related attacks is to generate the key with sufficient true randomness, which is the randomness contained in physical phenomenon such as the coin flipping and the dice rolling. In mission critical secure communication systems, the cryptographic keys are required to be generated by a true random number generator (TRNG), i.e. a device that collect random bits from random physical phenomenon.

The downside of true random number generator (TRNG) is its cost and portability. Coin flipping and dice rolling are too slow to generate random bits for practical applications. In order to generate a stream of truly random bits in high speed, dangerous radioactive materials or expensive quantum optical devices will be used [41]. Our current technology does not allow TRNG to be efficiently and economically implemented into our daily communication devices, such as the laptop computers and the mobile phones.

Even if the key is generated with sufficient entropy, i.e. true randomness, the adversary may crack the key through cryptanalysis attack to the cipher that uses the key. For example, Refs. [63] and [99] studies the methods to reveal the key from the encrypted texts by exploiting cipher vulnerabilities. A notable incidence of such key cracking attacks is the cryptanalysis to the RC4 cipher, which is widely used in wireless LAN security [104]. Because a vulnerable design of the RC4 cipher was standardized in the wired equivalent privacy (WEP) mode of wireless LAN security, adversary can crack the wireless key within several minutes using a laptop computer [135].

There is a fundamental conflict between the entropy requirement of passwords and the memorization capability of our brains. An interesting example of this conflict is demonstrated by Russian spies. In 2011, U.S. government cracked down a Russian spy network and captured computers that contain encrypted information. The Russian spies use a long password, a string of 27 random characters, for the encryption. Such a complex password is uncrackable with today's computing technology. However, this password is too complex to be accurately memorized and the Russian spies have to write down the password on a piece of paper. The U.S. agents found the password in a drawer and decrypted the information in the captured computer [72]. In general, if a cryptographic key can be memorized, it is likely vulnerable to key cracking attacks. If it is complex and hard to crack, it will require storage devices other than our brain, and therefore is threatened by key stealing attacks.

2.2.2 Key Stealing

Instead of cracking the key through mathematical analysis and computations, the adversary may steal the key by intruding the device where the key is stored. The adversary may install a Trojan program to seize the passwords and private keys stored on the user's computer [102]. The adversary may also install a keylogger program to record the passwords input by the user [125]. It is even more dangerous when an adversary invades the servers that stores many passwords and cryptographic keys. In 2011, more than 100 million users' account information including passwords, are released by hackers [130].

Mission-critical cryptographic keys may be stored in tamper-proof hardware. The hardware executes cryptographic algorithms inside and rejects any attempt to read the key value from outside the hardware. Even with such careful measure, the adversary may obtain the value of the key through side-channel attacks [37, 93].

Social engineering attacks are the most versatile type of key stealing attacks. The adversary lures the victim users to give out their keys by deceiving them [123] instead of stealing their keys using advanced computing techniques. A simple example of social engineering attacks is the email phishing attack [52]. The adversary sends an email to the user and requests for the user to login to a fake website. The web site is created by the adversary and when the user type in his password, the adversary obtains it. It is very difficult to protect the users from being deceived using current security technologies [98].

Social engineering attacks can be completely "legitimate". Adversary may exploit the inconsistency of the security policies among multiple organizations. From the view of each organization, the security policies within the organization are self-consistent. The combined effects of security policies across different organizations may become security vulnerabilities.

In August 2012, a hacker pretended as the victim user to contact amazon.com online support. The amazon.com support allowed the hacker to visit the victim user's pre-login web page, which contained the last four digits of the user's credit card number. In amazon.com's security policies, the last four digits of a user's credit card number only serve as a hint to the user. Knowing these four digits would not cause security vulnerabilities in amazon.com. However, apple.com use the last four digits of a user's credit card number as the proof of the user's identity. The hacker called apple.com to ask for a password reset with the four digits he retrieved from amazon.com. The apple.com support verified the last four digits of the credit card number and granted the hacker's request. The hacker is then owning the user's apple.com account password through a sequence of legitimate actions [50]. Although apple.com and amazon.com later changed their security policies after this hacking incident was reported, such attacks are hard to avoid in general. It is difficult to predict such type of vulnerabilities and coordinate multiple organizations to implement preventive measures.

The above key cracking and key stealing examples are just a small portion of publicly known attacks. Many key thefts are just left unnoticed. *With today's technology, it is extremely difficult, if not impossible, to provision an information security system that would never leaks its key to adversary.*

Chapter 3
Dynamic Secrets

Instead of seeking a solution that can safeguard cryptographic key forever and defend every known and unknown attack, we propose to quickly recover a stolen key to a secret, i.e. a piece of information that is unknown to adversary. If the recovery period is sufficiently short, the security threat of key thefts would be effectively mitigated.

At a first glance, such a solution seems to be intractable. Since it is practically difficult to perceive a key theft when it occurs, we may not be able to "quickly" react and replace the stolen key with a new key. However, we may take advantage of random factors naturally reside in communications and force the key stealing adversary to quickly "forget" his stolen key.

This chapter presents *dynamic secrets* as the tool to implement our proposed solution approach. Dynamic secrets are a series of hash values generated from secure communication traffic. As a secure communication between two users goes on, both users consistently and synchronously generate dynamic secrets by hashing the information exchanged between them. These hash values are then used to dynamically update a symmetric key between these two users. Because the key value is constantly changing, the symmetric key between users is so called a dynamic key.

Section 3.1 provides a preview for dynamic key's security unique security properties. Section 3.2 introduces a packet communication model in a noisy environment that is a representation of a large number of practical secure communication scenarios. We use this model to present the processes of dynamic secrets generation and dynamic key updates. The remaining sections of this chapter analyze the security features brought by dynamic secrets in the noisy communication model. Section 3.3 investigate the process that the true randomness in communications is converted into the randomness of the key through dynamic secrets. Section 3.4 studies the conditions and the rate of automatic stolen key recovery. Section 3.5 presents a false-alarm free detection method to detect the attacks that an adversary uses a duplicate key to impersonate as a legitimate user to send and receive packets.

S. Xiao et al., *Dynamic Secrets in Communication Security*,
DOI: 10.1007/978-1-4614-7831-7_3,
© Springer Science+Business Media New York 2014

3.1 Dynamic Key Preview

Dynamic key is much less vulnerable to key theft attacks than a traditional symmetric key. The frequently changing key value provides a strong shield against cryptanalysis attacks. Even if a dynamic key is short and does not have sufficient entropy, exhaustive search of key is no longer an effective attack. Each successful exhaustive search only reveals a small segment of the secure communication. In order to be able to constantly decrypt secure communications between two users, the attacker must repeat his exhaustive search efforts for every small communication segment. The excessive demand for computational resources makes the exhaustive search attack impractical.

Furthermore, the key value is associated with the secure communication traffic. The key value inherits true randomnesses from the users' generated messages which are exchanged in the secure communication and true randomness from the communication environment which shapes the communication traffic. The true randomness makes dynamic key truly unpredictable and uncrackable by any computational algorithm.

Suppose a strong adversary successfully obtains a dynamic key. The adversary is forced to maintain an everlasting and error-free eavesdropping effort on users' communication traffic since the moment he stole the key. Otherwise he will not be able to keep tracking on the value changes of dynamic key. Such an eavesdropping will substantially increase the adversary's costs of attack and risks of being exposed.

Moreover, random factors such as communication channel noises and imperfections in the adversary's receiver cannot be eliminated in practice. These random factors make perfect eavesdropping practically impossible. Once the adversary encounters decoding error or packet loss in his eavesdropping, he will not be able to calculate the value that the dynamic key updates to. After the key update next to adversary's information loss, the adversary's knowledge to the dynamic key will no longer be valid. From the users' perspective, the adversary "forget" the stolen key. The dynamic key automatically recovers to a secret after the key theft.

We could further allow the situation to be extremely favorable for the adversary. Suppose a mighty adversary steals a dynamic key and keeps an error-free eavesdropping on the users' communication. In this extreme setting, the adversary is able to passively eavesdrop and decrypt the secure communication. Unlike the traditional secure communication case, where an adversary with the key is hardly distinguishable from a legitimate user, it is impossible for an adversary with dynamic key to impersonate as a legitimate user to send or receive messages without being detected.

The adversary's fraudulent message changes the value of dynamic key in the user who receives the message. However, dynamic key of the user who is impersonated by the adversary remains unchanged. The discrepancy between dynamic keys is an undeniable evidence that reveals the existence of fraudulent message. Dynamic secrets and dynamic key guarantee that any type of active attacks will be accurately and promptly detected. Such a detection mechanism is fundamentally different to traditional intrusion detection mechanisms, which rely on auxiliary information like IP addresses or users' behavior statistics. Dynamic secrets based detection mechanism consumes negligible amount of computational resources and does not trigger any false positive alarm.

3.2 Dynamic Secrets in Noisy Packet Communications

We begin the detail, technical discussion of dynamic secrets and dynamic key by introducing three classic role players, Alice, Bob, and Eve. Alice and Bob are legitimate users. Eve is the adversary who can eavesdrop communications between Alice and Bob. Eve may also steal the key used to secure the communications between Alice and Bob.

3.2.1 Noisy Packet Communication Model

Alice wants to send a sequences of data packets to Bob in a noisy communication environment, such as when Alice and Bob use wireless communication. Because decoding errors and packet losses are unavoidable, Alice and Bob use packet retransmission mechanism, such as the Stop-and-Wait protocol [120], to ensure reliable packet delivery.

In Alice-Bob communications, each data packet carries a unique sequence number. The data packets are transmitted in increasing order of their sequence numbers. We further assume that each data packet has a retransmission flag bit in its packet header and the recipient Bob knows exactly whether a received data packet is delivered through retransmissions or not.

The adversary Eve is allowed to eavesdrop all communications between Alice and Bob. Because of the noisy environment, Eve also experiences unavoidable decoding errors and packet losses. It is worth clarifying that a packet loss for Eve means she misses the first transmission and all retransmissions of a packet. If Eve misses the first transmission of a packet but correctly receives at least one of its retransmissions, this packet will not be counted as Eve's packet loss.

Eve cannot request packet retransmissions to cover her packet losses. Otherwise Alice and Bob would notice Eve's existence. After a period of communication, all the data packets sent by Alice have been correctly received by Bob. Some of these data packets are not correctly received by Eve during her eavesdropping.

This model is a general representation of many practical packet communication scenarios, in which random factors in the communication environment ubiquitously affect both the users and the adversary. Figure 3.1 illustrates this noisy packet communication model.

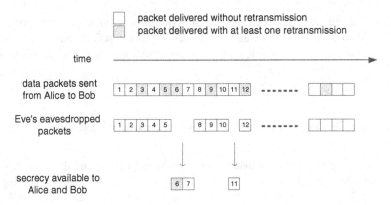

Fig. 3.1 A lossy packet communication model with the presence of the eavesdropping adversary

From Alice's and Bob's views, some data packets are correctly received by Bob without any retransmission. Other packets are delivered through multiple transmissions. As shown in Fig. 3.1, their communication is free of information loss.

From Eve's perspective, as Alice keeps sending packets to Bob, Eve's eavesdropping packet losses are unavoidable. On average, Eve may have fewer packet losses than Bob has. However, Bob can recover all of his packet losses by requesting packet retransmissions. Eve's packet losses are non-recoverable, such as packets 6, 7, and 11 in Fig. 3.1.

Consider the first 12 packets in the communication. Alice and Bob both have complete knowledge about all these packets, while Eve misses 3 of these packets. Specifically, packets 6, 7, and 11 contain information that Alice and Bob commonly know but Eve does not know. The information carried by these packets are the users' information advantage over the adversary. By definition, these information are users' shared secrets.

Moreover, the information advantage, i.e. users' shared secrets, is naturally established during the communication process. Alice and Bob do not have to pay additional computational or bandwidth cost to establish their shared secrets.

3.2.2 Dynamic Secrets and Dynamic Key

We illustrate the generation of dynamic secrets and the dynamic update of symmetric key by showing a modified secure packet communication process. Suppose Alice and Bob have a pre-shared, symmetric key $k(0)$. They use $k(0)$ to drive a symmetric cryptographic algorithm such as AES to encrypt and decrypt data packets.

In order to generate dynamic secrets, Alice and Bob each maintains a small packet buffer that can temporarily hold a number of data packets. Bob stores his received data packets in his packet buffer. Alice stores the packets that are confirmed to be

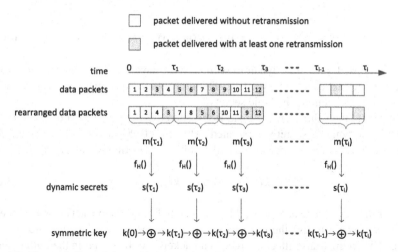

Fig. 3.2 The work flow of dynamic secrets generation and dynamic key update. $f_H(\cdot)$ is a hash function. $\oplus$ is the bitwise-XOR operator

received by Bob into her packet buffer. The example in Fig. 3.2 requires Alice's and Bob's packet buffers to be able to hold at least four packets.

For every 4 delivered packets, which have been stored in packet buffers, Alice and Bob generate a dynamic secret from these 4 packets and update their symmetric key once. Then Alice and Bob empty their packet buffers to accommodate next 4 delivered packets. As the communication goes on, Alice and Bob iteratively generate dynamic secrets and update their symmetric key. This iterative process is shown in Fig. 3.2. Each iteration can be decomposed into three steps: packet rearrangement, dynamic secret generation, and dynamic key update.

Whenever 4 data packets have been delivered and buffered, Alice and Bob rearrange these four packets according to the following rules.

1. packets are collected into two groups according to whether they are retransmitted or not with packets in each group in the order of their sequence numbers.
2. the group of retransmitted packets is placed after the group of non-retransmitted packets.

As shown in Fig. 3.2, packets 5, 6, 7, and 8 have been delivered and buffered by Alice and Bob. Packets 5 and 6 are retransmitted packets. Packets 7, and 8 do not encounter retransmission. Therefore, Alice and Bob firstly split these 4 packets into two groups: {5, 6} and {7, 8}. Then the retransmitted packet group {5, 6} is placed after the non-retransmitted packet group {7, 8}. The final order of packets after packet rearrangement is {7, 8, 5, 6} as shown in the figure.

We denote the times of the packet rearrangements as $\tau_1, \tau_2, \ldots$. At time τ_i, $i = 1, 2, \ldots$, both Alice and Bob construct a binary string $m(\tau_i)$ by concatenating the plain texts of the rearranged data packets, including the packet headers. Then Alice and Bob apply a cryptographic hash function $f_H(\cdot)$ on $m(\tau_i)$ to generate a hash value

$s(\tau_i)$.

$$s(\tau_i) = f_H(m(\tau_i)) \tag{3.1}$$

$f_H(\cdot)$ is chosen such that dynamic secrets $s(\tau_i)$ have the same binary length as $k(0)$, the pre-shared symmetric key. These dynamically generated hash values, $s(\tau_1), s(\tau_2), \ldots$, are the dynamic secrets.

We denote the value of Alice's and Bob's shared symmetric key at time t as $k(t)$. Whenever a dynamic secret is generated, Alice and Bob perform a key update. At time $\tau_i, i = 1, 2, \ldots$, the key is updated by bitwise-XORing it with $s(\tau_i)$.

$$k(\tau_i) = k(\tau_i^-) \oplus s(\tau_i) = k(\tau_{i-1}) \oplus s(\tau_i) \tag{3.2}$$

After the key update at time τ_i, Alice and Bob discard the packets stored in their buffers and begin to use the updated key $k(\tau_i)$ to encrypt and decrypt the next 4 packets. Then the transmitted and received packets will be stored in the buffers and then trigger the next round of dynamic secret generation and dynamic key update. The iterations occur once every 4 packets have been delivered throughout the communication process.

It is worth noting that the above defined dynamic secrets are generated from the data packets sent from Alice to Bob. The dynamically updated cryptographic key is used to encrypt and decrypt the data packets communicated along this direction. In practice, packet communications can be from Alice to Bob and from Bob to Alice, the data packets sent from Bob to Alice are used to generate another series of dynamic secrets which are independent to the dynamic secrets series generated from the packets sent from Alice to Bob. Therefore, the dynamic key used to encrypt and decrypt packet sent from Bob to Alice will be independent to the dynamic key used for the packet communication from Alice to Bob. Although both dynamic keys are symmetric and updated from the same pre-shared symmetric key, $k(0)$.

3.3 True Randomness at Low Cost

As presented in Chap. 2, the first major type of key safety risks is that adversary may be able to guess the key value, either through cryptanalysis to the key generation algorithm or by brute force search. The keys generated from a pseudo random number generator may be vulnerable to algorithmic attacks as well as the statistical inference attacks that target the limited truly randomness in the random seed used to generate the key [18, 124]. Preparing a large number of true random bits for the cryptographic key generation involves expensive hardware and may not be applicable for the mobile secure communication users [41].

In response to this type of key safety risks, dynamic secrets harvest harvest *true randomness* from the communication environment. This randomness is then conveyed into the cryptographic key through the dynamic key updates. The algorithmic attacks are no longer valid since the key value is affected by the physical randomness in the communication environment. In order to guess the value of a dynamic key, adversary needs to predict physical random fluctuations in the communication environment, which is practically impossible. Moreover, because dynamic secrets keep harvesting randomness and fueling information entropy into the key, brute force search would not work on dynamic key.

The true randomness is harvested in the packet rearrangement step. When Alice sends a packet to Bob, whether this packet needs to be retransmitted or not is determined by the instantaneous state of the communication channel. The retransmission information of this packet is associated with physical, true randomness in the channel. When a bunch of delivered packets are rearranged, the final packet order is determined by the retransmission information of these packets, which contains true randomness.

Different packet orders lead to different values of the generated dynamic secret. The values of dynamic secret contain true randomness. When a dynamic secret is XORed with the key in a key update, the updated key will inherit the true randomness from the dynamic secret. Figure 3.3 shows possible outcomes of the packet rearrangement when Alice and Bob rearrange every four delivered packets for their secure communications.

Suppose packet losses are independent events. Each packet has a probability $1 - p_l$ to be delivered without retransmission. We use Shannon entropy [30] to quantify the true randomness harvested by rearranging these data packets. Let $\mathcal{O}$ represent the set of all possible packet orders after rearrangement.

$$
\begin{aligned}
H(\textbf{harvested true randomness}) \\
= \sum_{o \in \mathcal{O}} -p(o) \log p(o) \\
= -3p_1 \log p_1 - 5p_2 \log p_2 - 3p_3 \log p_3 - (\sum_{i=0}^{4} p_i) \log (\sum_{i=0}^{4} p_i),
\end{aligned}
\tag{3.3}
$$

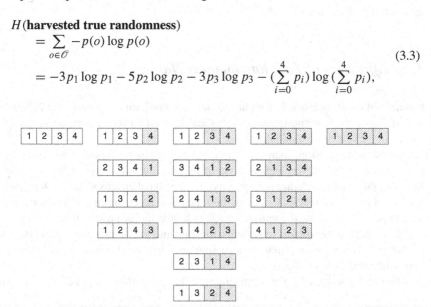

Fig. 3.3 Possible orders of 4 data packets after rearrangement. *White blocks* represent data packets delivered without retransmission. *Shadowed blocks* represent data packets delivered with retransmissions

in which the $\log(\cdot)$ function is base 2 and

$$p_0 = (1 - p_l)^4$$
$$p_1 = (1 - p_l)^3 p_l$$
$$p_2 = (1 - p_l)^2 p_l^2$$
$$p_3 = (1 - p_l) p_l^3$$
$$p_4 = p_l^4.$$

Let $p_l = 0.05$, which is typical in the wireless communication scenarios such as indoor wireless LAN communications and open field mobile wireless communications. Equation (6.35) shows that by rearranging 4 packets, 0.8756 bits of true randomness is harvested. If the packet delivery rate is 100 packets per second, then dynamic secrets can constantly harvest true randomness at a rate of 21.89 bits per second. In other words, dynamic secrets can harvest sufficient true randomness that is equivalent to the randomness contained in a hardware TRNG generated 128-bit symmetric key in every 6 seconds.

As long as the communication continues, true randomness from the physical communication environment is constantly harvested and used to fueling the key entropy. More importantly, the randomness harvest is based on the information already shared between Alice and Bob during the communication process. The cost for Alice and Bob to obtain the true randomness is practically zero, in contrast to the high cost associated with the use of TRNG.

3.3.1 Efficiency of True Randomness Harvest

It is natural to ask questions about how much true randomness is available to be harvested during the communication process and how efficiently the packet rearrangement can harvest these randomness.

For the first question, there are studies that suggest to exploit signal propagation delay and the received signal strength to generate truly random keys. In wireless communications, channel reciprocity and the signal envelopes are also possible sources of true randomness. However, these physical layer parameters are not conveniently accessible in most communication systems. In our proposed scheme, Alice and Bob rely on the packet retransmission information to rearrange packets. The packet retransmission information, as a source of true randomness, is available for both Alice and Bob at no cost.

Alice and Bob look for the retransmission flag bit in the packet header to obtain the packet retransmission information. Typically, the retransmission flag bit only takes one bit in the packet header. Alice and Bob can only distinguish a retransmitted packet from a non-retransmitted packet. They cannot distinguish a packet that is retransmitted once from a packet that is retransmitted more than once. Therefore, for

4 delivered packets, the true randomness available for harvesting is

$$H(\textbf{available true randomness}) = 4(-p_l \log p_l - (1 - p_l) \log (1 - p_l)). \quad (3.4)$$

When $p_l = 0.05$, there are 1.1456 bits of true randomness carried by the retransmission flag bit in these 4 packets.

According to Eq. (6.35) the rearrangement of 4 packets contains 0.8756 bits of true randomness. It is approximately 75 % of the true randomness available. A portion of the true randomness is lost during the packet rearrangement because some different packet retransmission patterns result in the same packet order outcome after the rearrangement.

The above computations can be generalized for the rearrangement of an arbitrary number of packets. The total true randomness available to be harvested for n delivered packets is

$$H(\textbf{available true randomness}) = n(-p_l \log p_l - (1 - p_l) \log (1 - p_l)). \quad (3.5)$$

Suppose Alice and Bob buffer the n delivered packets and rearrange them. $p_i = (1 - p_l)^{n-i} p_l^i$ for $i = 0, 2, \ldots, n$. We have

$$H(\textbf{harvested true randomness})$$
$$= \sum_{o \in \mathcal{O}} -p(o) \log p(o)$$
$$= -(\sum_{i=0}^{n} p_i) \log (\sum_{i=0}^{n} p_i) - \sum_{i=1}^{n-1} (\binom{n}{i} - 1) p_i \log p_i \ \textbf{bits}, \quad (3.6)$$

in which $\binom{n}{i}$ is the number of different combinations to choose i items from n items.

Figure 3.4 shows true randomness harvest efficiency as a function of the number of packets to rearrange, n and the probability of Bob's packet loss, p_l.

$$\textbf{harvest efficiency} = \frac{H(\textbf{harvested true randomness})}{H(\textbf{available true randomness})} \quad (3.7)$$

Alice and Bob can improve true randomness harvest efficiency by buffering more packets before packet rearrangement. A noisier communication environment is also more favorable for harvesting true randomness.

Let $H(p_l) = -p_l \log p_l - (1 - p_l) \log (1 - p_l)$. The number of truly random bits harvested per delivered packet can be expressed as

$$\textbf{\# of truly random bits per packet} = H(\textbf{harvested true randomness})/n$$
$$= H(p_l) * \textbf{harvest efficiency} \quad (3.8)$$

We can plot the curves for the number of truly random bits harvested per delivered packet by scaling the curves in Fig. 3.4. Both $H(p_l)$ and the harvest efficiency are less than 1 for all finite n. Therefore Alice and Bob cannot harvest more than one

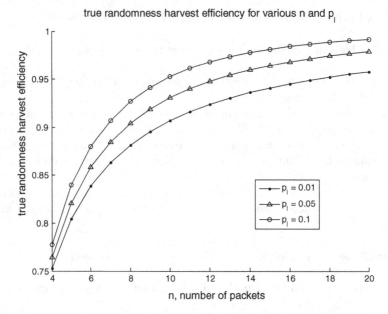

Fig. 3.4 True randomness harvest efficiency by packet rearrangement for different *n* and p_l combinations

truly random bit per delivered packet. This upper bound is apparent because each packet only has 1 flag bit to carry packet retransmission information.

The rearrangement of every *n* packets harvest a certain amount of true randomness. The *total* amount of available true randomness and the *total* amount of harvested true randomness both grow without an upper bound as the number of delivered packets increases. Dynamic secrets can convert a noisy communication channel into a rich source of true randomness.

3.4 Automatic Stolen Key Recovery

The traditional communication security paradigm is vulnerable to key theft attacks. Once an adversary obtains the key, cryptographic services such as authentication, encryption and keyed message integrity check become insecure. The compromised communication security can only be reinstated after a new key is issued to replace the stolen key.

Because of practical constraints such as cost and operational convenience, cryptographic key updates are typically scheduled at intervals of months or even years for secure communication systems. Once a cryptographic key is stolen, the time period between the key theft and the issuing of a new key to replace the stolen key would be very long. In this case, secure communication users would experience a long time gap without security. The adversary would have a long period of time to attack these systems using the stolen key.

The problem is further exacerbated when users rely on an existing key to establish a secure communication channel for key updates. If the existing key is stolen and known to the adversary, even a key update would not reset the communication security. The adversary can eavesdrop the secure key update process, which is authenticated and encrypted by the stolen key. He will be able to know the new key at the moment it is issued to the user.

Traditional communication security research devotes much efforts to develop techniques to strengthen key protection in order to prevent the adversary from obtaining the key. As discussed in Chap. 2, to protect a cryptographic key from being stolen is an extremely challenging task. It is practically impossible to eliminate all key theft vulnerabilities and construct a secure communication system that guarantees its key would not be known to adversary in any case.

Dynamic secrets provide a new way to mitigate the key theft problem. Instead of trying to prevent adversary from obtaining the key, the use of dynamic secrets allows an automatic recovery for the stolen key and the communication security when communication environment is noisy. If users communicate frequently, the automatic stolen key recovery would be prompt and severely limit the adversary's time to attack using the stolen key. Figure 3.5 demonstrates the process of automatic stolen key recovery.

As shown in Fig. 3.5, Alice keeps sending packets to Bob. Bob's packet losses are compensated by packet retransmissions. Suppose Eve steals the key at time t_{stolen}. At the moment of key theft, Eve obtains the value of $k(\tau_i)$.

Starting from t_{stolen}, Eve must keep eavesdropping the communication traffic between Alice and Bob. Otherwise at time τ_{i+1}, dynamic secrets will update the key to $k(\tau_{i+1})$ and Eve's knowledge of the key would be no longer useful for him.

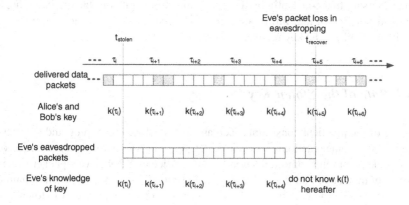

Fig. 3.5 As Alice and Bob continues their communication, the stolen key will recover when Eve encounters a packet loss

The communication environment is noisy, as Alice and Bob keep sending and receiving packets, Eve will inevitably encounter packet losses in eavesdropping. Suppose the timing of the Eve's first packet loss is between τ_{i+4} and τ_{i+5}, as shown in Fig. 3.5. The key update at time τ_{i+5} immediately restores the communication security because Eve cannot know $k(\tau_{i+5})$ and all the key values thereafter.

When Alice and Bob maintain a high speed communication in a noisy environment, such as a wireless LAN communication in a crowded office building, the stolen key can be restored to a secret within a fraction of a second after the key theft. Such a prompt recovery impairs Eve's capability to use the stolen key for malicious attacks, and thus defeats the purpose of key theft.

Environment noise is not the only factor that will result in Eve's packet losses. Signal interferences, artificial noises, and imperfection of Eve's receiver all contribute to her packet loss rate. Even in a wired communication environment such as the Internet, if Alice randomly pick multiple routes to send packets to Bob and Eve cannot eavesdrop all possible routes for all times, dynamic secrets will function properly and provide strong protection against key theft attacks.

Theoretically speaking, the probability that Eve maintains packet loss free eavesdropping for a long but finite period of time is not zero. However, in practice, such a small chance of "success" will certainly hurt Eve's incentive to steal the key.

From Alice's and Bob' point of view, they circumvent the difficult problem of detecting key thefts in time. When dynamic secrets are used, the recovery of stolen key is automatic. Alice and Bob do not have to be aware of key thefts.They simply keep communicating with each other and Eve's packet loss will eventually invalidate her knowledge of the key as if the key is self-recovering.

Dynamic secrets switch the roles of attacker and defender in the key theft problem. In traditional secure communication systems, the users have to constantly protect their key from being stolen. The longer a key has been used, the more likely it is stolen and endangers the communication security. When dynamic secrets are used, it is the adversary that constantly battle against all sorts of random factors in the hope of eavesdropping without information loss. The longer adversary holds the stolen key, the more likely the key recovers to a secret.

3.4.1 Rate of the Stolen Key Recovery

Because of the practical constraints such as the cost of eavesdropping and the presence of the channel randomnesses, it is not possible for Eve to correctly receive every packet from the communication between Alice and Bob. Eve's packet loss is practically inevitable and the stolen key will eventually be restored when dynamic secrets are used to update the key. It is interesting to study the speed of stolen key recovery.

The speed of stolen key recovery depends on the users' packet delivery rate, the adversary's packet loss rate, and the amount of information contained in each packet. To simplify the analysis, we assume the key is recovered whenever Eve completely misses one packet in her eavesdropping. This assumption is true for many practical communication scenarios. Take Internet Protocol version 4 (IPv4) packets as an example. Most IPv4 data packets contains more than 100 bytes of payload information. When Eve misses such a packet, she has to recover every bit in the payload in order to calculate the next key update. Such a recovery equivalents to predict more than 100 bytes of source data that Alice sends to Bob. It is harder than to make a random guess on the key, which typically only has 128 or 256 bits.

One may argue that Eve may reconstruct a lost packet by using semantic analysis and information from other packets that Eve correctly eavesdropped. This is theoretically possible but impractical in most cases. Moreover, as communication goes on, Eve would have indefinite number of packet losses, the chance to reconstruct all of them and keep tracking dynamic key updates is negligible.

Suppose the packet delivery rate is R when Alice sends packets to Bob. Eve has a packet loss rate $p_{l,Eve}$. This rate only accounts for "complete" packet losses, i.e. for packets that Eve missed their transmissions and all retransmissions. Assume packet losses are independent random events, we have

$$p(\textbf{Eve still knows key at time } t) = (1 - p_{l,Eve})^{Rt}. \qquad (3.9)$$

This probability decreases exponentially. The average time for the stolen key to be recovered is

$$\overline{T}_{recovery} = \frac{1}{p_{l,Eve}R}. \qquad (3.10)$$

Equations (3.9 and 3.10) show that increasing p_L and increasing R both help Alice and Bob to speed up the recovery of a stolen key. Alice and Bob can communicate at a higher rate or introduce artificial jamming to the communication environment to obtain a faster stolen key recovery.

Equation (3.10) gives a sufficiently short key recovery time in many practical packet communication scenarios. For example, in an indoor 802.11b network, the packet loss rate for a commercial-off-the-shelf (COTS) receiver typically ranges from 1 to 5 % [65]. If Eve uses a COTS receiver as well and Alice and Bob communicate at a low rate of 1 packets per second, a stolen key is expected to recover in roughly a minute. Eve may lower her packet loss rate using MIMO technology and a high-gain antenna. The hardware upgrade will make Eve's eavesdropping more noticeable in the indoor environment.

In wired communications, the channel noise could be low and may not produce a significant packet loss rate. However, the wired packet delivery rate can be as high as tens of thousands of packets per second. Such a high speed communication is also difficult to eavesdrop without any information loss. Cronin et, al. systematically study eavesdropping in both wireless and wired communications. They conclude that current eavesdropping technology cannot reliably avoid packet losses in both wireless and wired communication scenarios [31].

We may compute the speed of stolen key recovery for different models of Alice-Bob communication. For example, the packet delivery rate between Alice and Bob could be a variable instead of a constant R. Eve's packet loss model can be constructed in many different ways. Modeling Eve's packet loss in bursts may be more realistic than assuming independent packet losses. The modeling differences would not affect the eventuality of Eve's packet loss and therefore would not invalidate the automatic stolen key recovery by dynamic secrets.

Assuming that Eve uses a receiver with reasonable technology and cost constraints, the speed of stolen key recovery can be expressed in terms of users' packet delivery rate and the environmental noise level, both of which can be measured. Therefore, the use of dynamic secrets provides a method of objectively evaluating the resilience level of a secure communication system against key theft attacks. Such an objective and quantitative assessment would improve the users' confidence in the security of their secure communication system. On the other hand, the users can control the key theft resilience level of their security by altering the packet delivery rate and the noise level in their communication environment.

3.5 Inherent Detection to Impersonation Attack

We further allow Eve to use the stolen key to attack before it is recovered. In this case, Eve would be able to decrypt her eavesdropped communication. However, dynamic secrets ensure that Eve cannot launch an impersonation attack without being detected.

An impersonation attack is that adversary disguises as a legitimate user to actively communicate with another user. In order to launch an impersonation attack, the adversary needs to obtain the cryptographic key. Therefore, users cannot detect the impersonation attacks using key verification or key based authentication schemes.

Traditional impersonation attack detection methods depend on auxiliary information such as user's IP address or hardware fingerprints to distinguish a legitimate user from an adversary [100, 116, 134, 143]. These methods could be defeated by an adversary who carefully fakes the auxiliary information. It is possible to detect impersonation attacks behavior pattern classification [108, 122, 133]. However, the behavior pattern classification relies on statistical methods that are not only computationally intensive, but also characterized by non-negligible false alarm probabilities.

The use of dynamic secrets enables a lightweight and accurate approach to detect impersonation attacks. When an adversary impersonates a user to communicate with another user using a stolen key, every bit sent from the adversary affects the key updates of the receiving user, while the impersonated user's key remains unchanged. Therefore, as soon as these two legitimate users communicate after the impersonation attack, they will immediately notice the discrepancy between their dynamic keys and become aware of the impersonation attack. This detection method does not require any computation. It is also free of false alarms.

Suppose Alice is sending packets to Bob. Figures 3.6 and 3.7 demonstrate the detection processes for two elementary types of impersonation attacks. In each type of attacks, Eve only impersonates as either Alice or Bob, but not both.

Fig. 3.6 Using dynamic key discrepancy to detect the impersonation attack that Eve impersonates as Alice

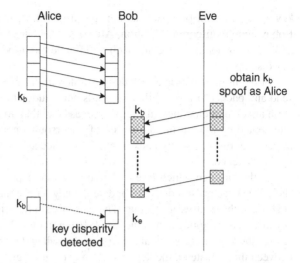

Fig. 3.7 Using dynamic key discrepancy to detect the impersonation attack that Eve impersonates as Bob

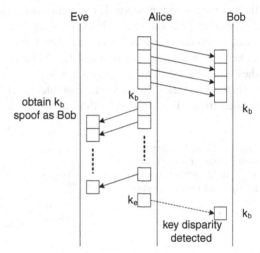

As illustrated in Fig. 3.6, Eve obtains key k_b and then uses this key to encrypt and send fraudulent packets to Bob. Because k_b is legitimate, Bob is deceived and accepts these fraudulent data packets. When Alice and Bob resume their packet transmissions, Bob's key has been updated to k_e and Alice still uses k_b as her key. The key discrepancy reveals the existence of the impersonation attack. Moreover, because Alice's key remains unchanged from the beginning of the impersonation attack to the detection of the attack, the timing of the attack can be precisely determined. The fraudulent packets can be easily identified.

Another type of impersonation attacks happens when Eve intercepts the packets sent from Alice and spoofs as Bob to acknowledge Alice for these packets. Eve does not actively sends any data packets to Alice. As shown in Fig. 3.7, Eve uses the stolen key k_b to intercept two packets that sent from Alice. In this case, when Alice and Bob resume their communication, Alice's key updates to k_e and Bob's key remains as k_b. Subsequently, Alice and Bob will notice the impersonation attack from their key discrepancy.

For the impersonation attack demonstrated in Fig. 3.6, we assume Alice does not send any packet to Bob while Eve sends her fraudulent packets. For the impersonation attack demonstrated in Fig. 3.7, we assume that all the packets from Alice are intercepted by Eve during the period of impersonation attack. Once Eve begins her impersonation attack, any packet sent from Alice and received by Bob will uncover Eve's existence.

For the cases in which Eve only sends or intercepts a small number of packets such that Alice or Bob does not update the key during the impersonation attack, the analysis to these cases are similar. There is only slight differences in the timing when key discrepancy appears than the cases shown in Figs. 3.6 and 3.7. Alice and Bob do not see the key discrepancy at the first packet when they resume the communication between them. Instead, the key discrepancy occurs right after the first key update in the resumed communication. If Eve impersonates as Alice to send fraudulent packets to Bob, then Bob will update his key earlier than Alice. If Eve impersonates as Bob to intercepts data packets from Alice, then Alice will update her key earlier than Bob. Either way, when Alice or Bob updates the key, the key discrepancy will appear and they will detect the impersonation attack.

In the worst case, Eve may send fraudulent packets to Bob and intercepts legitimate packets from Alice at the same time, as shown in Fig. 3.8. Eve intercepts all packets

Fig. 3.8 Using key disparity to detect the impersonation attack that Eve impersonates as Alice and Bob at the same time

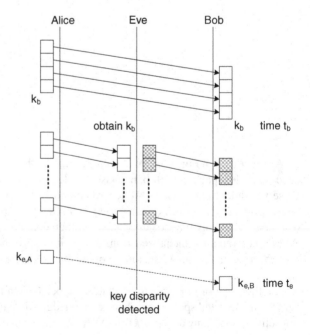

from Alice. She may forward some of the legitimate packets to Bob. She can also generate and send fraudulent packets to Bob. Unless Eve forwards all packets from Alice to Bob and does not inject any fraudulent packet, Eve cannot ensure Alice and Bob to have identical keys after her impersonation attack. Any manipulation Eve does to the Alice-Bob communication will cause dynamic key discrepancy between Alice and Bob and expose her existence.

3.5.1 Detect Impersonation Attack Without False Alarm

False alarms are the main drawback of traditional impersonation attack detection methods. There are two types of false alarms: false positive and false negative. The false positive corresponds to the detection method reporting an impersonation attack incident while there is none. For example, a legitimate user attempts to login from a public computer but the system treats him as an adversary and blocks his login attempt. The false negative is that the adversary successfully impersonates a user and the detection method fails to detect the attack.

Eliminate False Positives

The impersonation attack detection method based on dynamic secrets can avoid false positives by using a reliable transmission protocol that has the following properties.

1. each packet has a unique sequence number.
2. packet delivery is guaranteed by packet retransmissions.
3. both users are aware of whether a packet is delivered or not.
4. both users are aware of whether a packet is retransmitted or not.

One example of such a protocol is the Stop-and-Wait (SW) protocol with non-repetitive packet sequence numbers and a retransmission flag bit in each packet. Chapter 4 discusses the SW protocol in detail.

Because packet delivery is reliable and observed by both users, Alice and Bob will always generate identical dynamic secrets synchronously and maintain identical keys. Secure communication does not cause key discrepancy, and therefore false positives are eliminated.

Conditions for False Negatives

False negative occurs when the an impersonation attack occurs without being detected. When dynamic secrets are in use, Eve's modification to the packet flow between Alice and Bob will be firm evidence of her impersonation attack. The conditions for false negatives to occur are analyzed in the following scenario.

Alice sends packets to Bob and they update their key for every n delivered packets. At time t_b, Alice and Bob update their key to k_b. Suppose Eve obtains k_b and launches her impersonation attack as shown in Fig. 3.8.

The impersonation attack ends at time $t_e > t_b$ when Eve stops manipulating the communication between Alice and Bob. Bob receives packets sent from Alice directly. At this moment, Alice's key is denoted as $k_{e,A}$. The packets stored in her packet buffer is denoted as Ψ_A. Bob's key is denoted as $k_{e,B}$ and his buffered packets are denoted as Ψ_B. If $k_{e,A} = k_{e,B}$ and $\Psi_A = \Psi_B$, Alice and Bob will not be able to detect the existence of Eve and a false negative occurs.

Suppose Alice has m_A number of key updates during the time period $[t_b, t_e]$. Bob has m_B number of key updates during the same time period. The $m_A n$ packets used by Alice to update her key are collected into m_A groups, denoted as

$$P_i^A = \{p_{(i-1)n+1}^A, \ldots, p_{in}^A\} \tag{3.11}$$

for $i = 1, 2, \ldots, m_A$, where p_i for $i = 1, 2, \ldots, m_A n$ represents the $m_A n$ packets sent from Alice. Similarly, for Bob we have

$$P_i^B = \{p_{(i-1)n+1}^B, \ldots, p_{in}^B\} \tag{3.12}$$

for $i = 1, 2, \ldots, m_B$.

Ψ_A contains packets Alice sends after packet $p_{m_A n}$ and before time t_e. Ψ_B contains packets Bob receives after packet $p_{m_B n}$ and before time t_e. Therefore, Eve can ensure $\Psi_A = \Psi_B$ by forwarding all packets sent from Alice after packet $p_{m_A n}$ to Bob without adding any self-generated fraudulent packets.

Let $C(P_i^A)$ represent the concatenated binary string of the packets in set P_i^A, i.e.

$$C(P_i^A) = p_{(i-1)n+1}^A \| p_{(i-1)n+2}^A \| \cdots \| p_{in}^A, \tag{3.13}$$

where $\|$ denotes concatenation of binary strings. Similarly, we have $C(P_i^B)$ for Bob as

$$C(P_i^B) = p_{(i-1)n+1}^B \| p_{(i-1)n+2}^B \| \cdots \| p_{in}^B. \tag{3.14}$$

Therefore, we have

$$k_{e,A} = k_b \oplus f_H(C(P_1^A)) \oplus f_H(C(P_2^A)) \oplus \cdots \oplus f_H(C(P_{m_A}^A)) \tag{3.15}$$

and

$$k_{e,B} = k_b \oplus f_H(C(P_1^B)) \oplus f_H(C(P_2^B)) \oplus \cdots \oplus f_H(C(P_{m_B}^B)). \tag{3.16}$$

As a reminder of notations, $\oplus$ is the bitwise-XOR operator and $f_H(\cdot)$ is a cryptographic hash function.

$k_{e,A} = k_{e,B}$ is a necessary condition for Eve to hide her existence. Therefore, Eve must carefully manipulates the packet flow between Alice and Bob to have

$$\bigoplus_{i=1}^{m_A} f_H(C(P_i^A)) = \bigoplus_{i=1}^{m_B} f_H(C(P_i^B)) \qquad (3.17)$$

Eve cannot change what Alice sends. As a result, Eve cannot change the value on the left side of Eq. (3.17). Eve can decide what Bob receives and change $C(P_i^B)$ for $i = 1, 2, \ldots, m_B$ to hold Eq. (3.17) true. However, the cryptographic hash function $f_H(\cdot)$ is resistant to pre-image attacks, i.e. it is not feasible for Eve to find a binary string y for a given x such that $f_H(y) = x$. Eve does not have control on the hash values, $f_H(C(P_1^B)), f_H(C(P_2^B)), \ldots, f_H(C(P_{m_B}^B))$ and cannot manipulate these hash values to have their XORed result equal to $k_{e,A} = \bigoplus_{i=1}^{m_A} f_H(C(P_i^A))$.

Eve wants to be able to send fraudulent packets to Bob and satisfy Eq. (3.17) at the same time. The only option left for him is to forward all the packets sent from Alice to Bob and sends two identical sets of fraudulent packets such the hash values of these two sets of fraudulent packets cancel each other in the XOR operation.

In this case, we have $m_B \geq m_A$ because Bob will receive all packets sent from Alice plus fraudulent packets generated by Eve. Eve divide the sets of Bob's received packets, P_i^B for $i = 1, 2, \ldots, m_B$, into two groups. One group is an identical copy of P_i^A for $i = 1, 2, \ldots, m_A$, denoted as $P_{A(i)}^B$ for $i = 1, 2, \ldots, m_A$. The XORed result of the hash values on this group of packet sets equals to $k_{e,A}$, i.e.

$$\bigoplus_{i=1}^{m_A} f_H(C(P_{A(i)}^B)) = \bigoplus_{i=1}^{m_A} f_H(C(P_i^A)) = k_{e,A}. \qquad (3.18)$$

Another group is denoted as $P_{B(i)}^B$ for $i = 1, 2, \ldots, m_B - m_A$. The XORed hash value of this group of packet sets should be zero, i.e.

$$\bigoplus_{i=1}^{m_B - m_A} f_H(C(P_{B(i)}^B)) = 0. \qquad (3.19)$$

When Eqs. (3.18 and 3.19) stand at the same time, Alice and Bob will not see a key discrepancy because

$$
\begin{aligned}
k_{e,B} &= k_b \oplus \bigoplus_{i=1}^{m_A} f_H(C(P_{A(i)}^B)) \oplus \bigoplus_{i=1}^{m_B - m_A} f_H(C(P_{B(i)}^B)) \\
&= k_b \oplus \bigoplus_{i=1}^{m_A} f_H(C(P_i^A)) \oplus 0 \\
&= k_{e,A}.
\end{aligned}
\qquad (3.20)
$$

Eve would escape from being detected.

When $m_B > m_A$, the only feasible way for Eve to produce a zero for the XORed result in Eq. (3.19) is to send duplicate sets of packets and let their hash values cancel each other. This strategy will not succeed because the packet header contain a unique packet sequence number, which is a common practice. Bob will not place two packets with the identical sequence numbers into his packet buffer.

It is worth noting that the sequence number field in the packet header contains a limit number of bits and cannot increase without a bound. The sequence number will restart from 0 after it reaches the maximum presentable value. However, the period that the sequence number repeats itself is often large. As an example, the sequence number field for an IPv4 packet has 32 bits which means the maximum allowed sequence number could be $2^{32} - 1$. Such a value is often much larger than n, the number of packets buffered before each dynamic key update. In practice, within a key update period, the sequence number of a data packet is always unique.

Equation (3.19) can only be satisfied for one case, in which $m_B = m_A$. Eve acts as a passive relay device during the time period $[t_b, t_e]$. She forwards all packets from Alice to Bob without sending any fraudulent packets. In this case, Eve does not send any fraudulent packet and therefore does not launch an impersonation attack.

3.5.2 Man-in-the-Middle Attack

We want to raise the case of the Man-in-the-Middle (MITM) attack at the end of this chapter. The MITM adversary stands in the middle between two users and has the complete control of the communication channel between the users for an everlasting period of time.

Suppose Eve behaves as an MITM adversary and obtains the cryptographic keys, both the symmetric keys and the asymmetric keys, used by Alice and Bob. Eve also has an omniscient receiver that is capable of intercept all packets without any packet loss. All data packets sent from Alice are intercepted. Every data packet received by Bob are actually sent by Eve.

Eve can impersonate as Alice in front of Bob and impersonate as Bob in front of Alice at the same time. Alice and Bob cannot detect the presence of Eve because their communication is effectively cut off. They never have a chance to detect the attack by testing whether their dynamic keys are identical or not.

Indeed, an MITM adversary with cryptographic keys on hand cannot be defeated. It is impossible to maintain the communication security, with or without the help from dynamic secrets, when such a mighty adversary presents. The impossibility result has been proven by Maurer [83, 84].

We do not extend the discussion of MITM adversary here because Chap. 6 will focus on theoretic analysis including various adversarial models and present an intuitive explanation on why an everlasting MITM adversary with a complete knowledge of users' cryptographic keys is impossible to defend.

Except the everlasting MITM adversary with the key, dynamic secrets provide a method to detect the impersonation attack without false alarm. Every bit manipulated by the adversary becomes the evidence against him. Moreover, this method requires no computing and therefore is applicable to a large variety of secure communication systems.

Chapter 4
Dynamic Wireless Security

Dynamic secrets can make use of random factors in communications for security. On one side, dynamic secrets harvest true randomness and shield cryptographic keys from cryptanalysis attacks. On the other side, the randomness in communication environment help automatically reinstate communication security from key theft attacks.

Because of its random nature, wireless communication is a favorable environment for the idea of dynamic secrets. Noises, interferences, and many other random factors in wireless signal transmissions are natural sources of true randomness to be harvested by users. These random factors also produce unavoidable decoding errors and packet losses for the adversary.

This chapter takes wireless LAN as an example to demonstrate a simple dynamic secrets implementation in a practical secure communication system. Interestingly, the standardized link layer retransmission protocol of the wireless LAN allows a highly efficient implementation for dynamic secrets. Alice and Bob can synchronize their key updates without consuming any bandwidth.

Based on this implementation, we conduct a series of proof-of-concept experiments in an office wireless LAN environment. The implementation cost and the security performance of dynamic secrets are measured and discussed. The experimental results are quite encouraging and suggest a new way to deploy and operate a secure office wireless LAN.

Section 4.1 introduces a retransmission protocol of the Stop-and-Wait (SW) protocol that is commonly used at the wireless LAN link layer. Section 4.2 presents the algorithms that exploit characteristics of SW protocol to generate dynamic secrets and update dynamic key. Section 4.3 presents a prototype test platform that implements dynamic secrets and a series of experimental results using this test platform. Section 4.4 discusses the implementation of dynamic secrets in other layers of the secure wireless communication systems.

S. Xiao et al., *Dynamic Secrets in Communication Security*,
DOI: 10.1007/978-1-4614-7831-7_4,
© Springer Science+Business Media New York 2014

4.1 Stop-and-Wait (SW) Protocol

The transmission unit in link layer is frame, just like packets in the network layer. Our algorithms proposed in this chapter work with wireless LAN link layer mechanism and operate on wireless LAN frames. More specifically, our algorithm work with the Stop-and-Wait (SW) protocol, the reliable transmission protocol specified in the 802.11 standard [51] to retransmit frames corrupted by the noise in the wireless channel and to ensure the reliable delivery of frames.

Suppose Alice is sending data frames to Bob and they agree to use SW protocol for reliable frame delivery. Alice sends one frame at a time and stops and waits to receive an acknowledgment frame before sending the next frame. Bob returns an acknowledgment frame as soon as he correctly receives a data frame.

There are two possibilities that may prevent Alice from receiving an expected acknowledgment frame. Alice's data frame might be corrupted by wireless randomness and Bob does not reply with an acknowledgment frame. Even if Bob correctly receives a data frame and sends an acknowledgment frame back, the acknowledgment frame could be lost in wireless transmission as well. Either way, Alice may not receive an acknowledgment frame after send a data frame.

Alice does not wait forever; she starts a timer after sending each data frame. If she has not received the acknowledgment frame before a preset period of time, the timer would expire and signal a timeout. Alice would then resend the data frame, reset the timer and wait for the acknowledgment frame again. Alice only proceeds to send the next data frame when the current sending data frame has been confirmed to be correctly received by Bob.

The wireless LAN implementation of SW protocol has extra augments. Each frame carries a unique sequence number and a flag bit to mark whether this frame is in its first transmission or has been retransmitted. Figure 4.1 shows a typical wireless LAN frame transmission process.

Alice sends a data frame, m_1, with sequence number 1, to Bob. Bob correctly receives m_1 and immediately sends an acknowledge frame back to Alice. Alice stops and waits for the corresponding acknowledgment frame after transmitting m_1. She correctly receives the acknowledgment frame for m_1 and proceeds to send a new frame, m_2.

The first transmission of frame m_2 is lost in air because of the randomness in the wireless channel. Bob does not return an acknowledgment frame for the first transmission of m_2. Alice waits until a timeout occurs. She then marks m_2 as a retransmitted frame and resends it. This time Bob successfully receives m_2 and properly acknowledges it.

Alice receives the acknowledgment frame for m_2 and sends m_3. Bob receives the first transmission of m_3. However, Bob's acknowledgment frame for m_3 is corrupted by wireless randomness. Alice could not confirm the delivery of m_3. Therefore Alice resends m_3 after a timeout. Once Alice receives the corresponding acknowledgment frame of m_3, she then sends m_4. Frames m_4 and m_5 are delivered without retransmissions. Their acknowledgment frames also arrive Alice without problem.

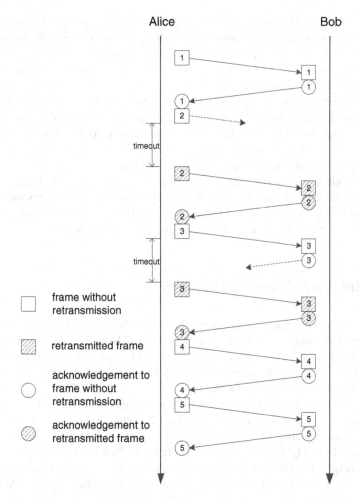

Fig. 4.1 An example of wireless LAN link layer frame transmission process using SW protocol with sequence number and retransmission flag bit in every frame

As shown in Fig. 4.1, SW protocol prevents Alice from sending a new data frame until both Alice and Bob both are aware of the successful delivery of the frame.

4.2 Dynamic Secrets in the Wireless LAN Link Layer

As discussed in Chap. 3, the implementation of dynamic secrets requires three steps: the rearrangement of delivered packets (frames), dynamic secrets generation, and dynamic key updates. The second and third steps only involve local operations, i.e. Alice and Bob could generate dynamic secrets and update the key based on their own data. However, the packet (frame) rearrangement step requires Alice and Bob to have common understanding of whether a packet (frame) has been retransmitted or not. Furthermore, Alice and Bob must be synchronous when starting to perform these three steps and keep their dynamic keys identical throughout their communications.

SW protocol has two characteristics which are helpful for the implementation of dynamic secrets. Firstly, SW protocol requires Alice stop and wait for the delivery of a packet (frame) before send the next packet (frame). This requirement ensures the synchronization of dynamic key updates. At the moment that Alice receives an awaited acknowledgment packet (frame), Alice and Bob have an identical set of buffered packets (frames). Dynamic key updates at this moment will yield same key value for both Alice and Bob. Secondly, SW protocol allow Alice and Bob to infer whether a delivered packet (frame) has been retransmitted or not with their local information. The process of identifying the packet (frame) retransmission status is named as Automatic Frame Classification (AFC) in this monograph.

4.2.1 Automatic Frame Classification

In order to perform frame rearrangement, Alice and Bob need to classify delivered frames into two groups: One-Time Frames (OTFs) and Non-One-Time Frames (non-OTFs). An OTF is a frame successfully delivered to Bob without retransmission. A non-OTFs is a frame that experiences retransmissions before Alice receives the corresponding acknowledgment frame.

Figure 4.2 reproduces the frame transmission process in Fig. 4.1 and marks the classification of each frame. As shown in Fig. 4.2, frames m_1, m_4, and m_5 are OTFs. Frames m_2 and m_3 are non-OTFs.

We introduce Algorithms 1 and 2 for Alice and Bob to classify their frames. With the help of the SW protocol, these algorithms work with only locally collected information. Alice and Bob can classify frames without exchanging any additional information. We call these algorithms as Automatic Frame Classification (AFC) algorithms. In our proposed scenario, Alice sends data frames to Bob. Therefore, Algorithm 1 is used by Alice. Algorithm 2 is used by Bob. We focus on unidirectional data communication in this chapter.

The following notations are used to present AFC algorithms. Alice and Bob use four frame buffers to temporarily hold a number of delivered frames: $\Psi_{s,1}$, $\Psi_{s,2}$, $\Psi_{r,1}$, and $\Psi_{r,2}$. $\Psi_{s,1}$ stores OTFs for Alice. $\Psi_{s,2}$ holds Alice's non-OTFs. $\Psi_{r,1}$ and $\Psi_{r,2}$ are used to store Bob's OTF and non-OTF respectively.

In each wireless LAN frame, its frame header contains two fields: a retransmission flag bit and a sequence number field. We use postfix .retran and .seq to denote these two fields in the frame, i.e. for a frame m, $m.retran$ denotes the value of its retransmission flag bit and $m.seq$ denotes its sequence number.

As specified in 802.11 standard, a wireless LAN frame has 12 bits to store sequence number. The field is sufficiently long for our dynamic secrets implementation to ensure every buffered data frame has a unique sequence number. In the following

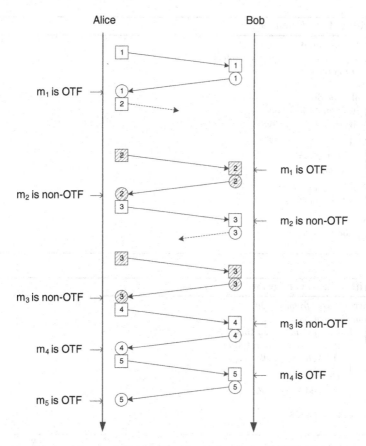

Fig. 4.2 An example of Automatic Frame Classification in wireless LAN link layer

discussions, we use m_i to denote the ith frame in our example scenario. It is not necessarily true that $m_i.seq = i$.

Alice can classify OTFs and non-OTFs in a straightforward manner, as specified in Algorithm 1. When Alice receives the acknowledgment frame for a data frame m_i after its first transmission and before a timeout, she identifies m_i as an OTF and placed into the OTF buffer $\Psi_{s,1}$. If Alice transmits m_i more than once, she classifies m_i as a non-OTF. Alice places a non-OTF m_i into $\Psi_{s,2}$ when the corresponding acknowledgment frame is received. After adding m_i into $\Psi_{s,2}$, Alice proceeds to send m_{i+1}.

Bob classifies a previously received frame when a new frame arrives as described in Algorithm 2. Let m_i be the most recently received frame, and m_{i-1} be the frame received prior to m_i. If m_i is a new data frame that has a different sequence number to m_{i-1}, Bob would know that m_{i-1}'s acknowledgment has been received by Alice and that m_{i-1} has been classified at Alice's side. Bob then check $m_{i-1}.retran$, the

---

Algorithm 1: AFC sender

---

1 **foreach** *frame* m_i **do**
2   set $m_i.seq = m_i$'s sequence number;
3   set $m_i.retran = 0$;
4   send m_i;
5   **while** *true* **do**
6   wait on ACK or time out;
7   **if** *ACK received* **then**
8   Jump out of loop;
9         $m_i.retran = 1$;
10   send m_i;
11   **if** $m_i.retran = 0$ **then**
12   Add m_i to $\Psi_{s,1}$;
13   **else**
14   Add m_i to $\Psi_{s,2}$;

Algorithm 2: AFC receiver

---

1 **foreach** *received frame* m_i **do**
2   **if** m_i *integrity check pass* **then**
3   **if** $m_i.seq \neq m_{i-1}.seq$ **then**
4   **if** $m_{i-1}.retran == 0$ **then**
5   Add m_{i-1} to $\Psi_{r,1}$;
6   **else**
7   Add m_{i-1} to $\Psi_{r,2}$;
8   send m_i's ACK;

---

retransmission flag bit field of m_{i-1}. If $m_{i-1}.retran = 0$, Bob adds m_{i-1} to $\Psi_{r,1}$. Otherwise, Bob adds m_{i-1} to $\Psi_{r,2}$.

Alice and Bob only use their local information to classify frames. Their classification results are identical. $\Psi_{s,1}$ and $\Psi_{r,1}$, $\Psi_{s,2}$ and $\Psi_{r,2}$ are naturally synchronized by the SW protocol. The AFC algorithms guarantee that $\Psi_{s,1} = \Psi_{r,1}$ and $\Psi_{s,2} = \Psi_{r,2}$ whenever Bob adds a frame to $\Psi_{r,1}$ or $\Psi_{r,2}$. Figure 4.2 marks the timing that Alice and Bob classify frames. Readers may follow the figure to check how these two algorithms work to classify OTFs and non-OTFs.

In $\Psi_{s,1}$ and $\Psi_{r,1}$, OTFs are placed in the ascending order of their sequence numbers. In $\Psi_{s,2}$ and $\Psi_{r,2}$, non-OTFs are ordered in the same way. Alice's frame rearrangement result is the string by concatenating $\Psi_{s,1}$ and $\Psi_{s,2}$, denoted as $\Psi_{s,1}||\Psi_{s,2}$. Bob rearrange his buffered data frames and obtain $\Psi_{r,1}||\Psi_{r,2}$.

When Alice and Bob perform frame rearrangement, they use the same sets of classified data frames. Therefore we can use the following notation to simplify the discussion.

$$\begin{cases} \Psi_1 = \Psi_{s,1} = \Psi_{r,1} \\ \Psi_2 = \Psi_{s,2} = \Psi_{r,2} \end{cases} \tag{4.1}$$

The string used to generate a dynamic secret is denoted as $\Psi_1 || \Psi_2$.

4.2.2 Dynamic Secrets Generation

Alice and Bob classify and buffer every delivered data frame on the fly. After collecting a number of data frames, they synchronously rearrange buffered frames, generate a dynamic secret, and update their cryptographic key. The timing for these operations is controlled by a threshold number n_{ts}, which is agreed by Alice and Bob before communication.

Algorithm 3 is used by both Alice and Bob to generate dynamic secrets.

Algorithm 3: Dynamic secret generation

---

1 **if** *at time t, either* $|\Psi_1| = n_{ts}$ *or* $|\Psi_2| = n_{ts}$ **then**
2     $s(t) = f_H(\Psi_1 || \Psi_2)$;
3   set $\Psi_1 = \Psi_2 = \phi$;

Algorithm 3 is executed when either Ψ_1 or Ψ_2 is filled up with n_{ts} frames. Therefore, a dynamic secret is generated from at least n_{ts} and at most $2n_{ts} - 1$ delivered and buffered data frames. The timing and the exact number of frames used in dynamic secrets generation is determined by the physical randomness in wireless communications. As a result, the outcome of Algorithm 3 is affected by true randomness from wireless environment. Even if Alice keeps sending data frames with the same payload, Alice and Bob would generate dynamic secrets with truly random and unpredictable values.

After generating a dynamic secret, buffers $\Psi_{s,1}$, $\Psi_{s,2}$, $\Psi_{r,1}$, and $\Psi_{r,2}$ are emptied. As communication continues, new frames would fill in these buffers and repetitively trigger Algorithm 3. A sequence of dynamic secrets is generated along with the secure wireless LAN communication. If the information contents contained in different data frames could be roughly viewed as independent to each other, the values of different dynamic secrets would be independent.

4.2.3 Dynamic Key Updates

Suppose at time t, a dynamic secrets, $s(t)$, is generated. Alice and Bob would update the symmetric key, $k(t)$, using $s(t)$ immediately after its generation. We use the following equation to define the dynamic key update in Chap. 3.

$$k(t) = k(t^-) \oplus s(t). \tag{4.2}$$

$k(t)$ could be updated to any key value from 0 to $2^{|k(t)|} - 1$. $|k(t)|$ is the binary length of $k(t)$.

In practice, not all possible key values could be used in cryptographic algorithms and protocols. For a key based cryptographic algorithm, some key values may cause the algorithm to behave in an undesirable way. These key values are known as weak keys. For example, a string of all zeros, **0x0000000000000000**, is a known weak key for the Data Encryption Standard (DES) [12]. Weak keys are often specified in the algorithm specification documents. Advances in cryptographic research can also uncover new weak keys.

In a practical implementation, the compatibility between the updated key and the encryption and decryption algorithms used in the wireless LAN communications needs to be considered. Therefore we use Algorithm 4 for dynamic key updates. The algorithm first stores the potential key update value in a variable k_{tmp} and test if k_{tmp} is a weak key for cryptographic algorithms and protocols based on $k(t)$. If k_{tmp} coincides with a known weak key value, the algorithm would skip a key update. $k(t)$ remains as its original value and works in the secure communication until the next effective key update.

Algorithm 4: Dynamic key update

---

1 **if** *a new dynamic secret $s(t)$ is generated* **then**
2 $\quad$ $k_{tmp} \Leftarrow k(t) \oplus s(t)$
3 $\quad$ **if** k_{tmp} *is not weak* **then**
4 $\quad\quad$ $k(t) \Leftarrow k_{tmp}$

Weak keys rarely appear in practice. Typically, only a small number of specific values will be marked as weak keys. For example, only 68 values are known as weak keys in 2^{56} possible key values for DES. There is no publicly known weak key for the AES Rijndael cipher when this monograph is written. In our proof-of-concept experiments, we have sent and received millions of wireless LAN frames with randomly generated payloads. hundreds of thousands of dynamic key updates have been triggered. We did not observe any weak key occurrence that requires Alice and Bob to skip a dynamic key update.

There is another compatibility issue that needs to be considered in dynamic key updates. If a poorly designed encryption algorithm is deployed in the wireless LAN, the commutativity of XOR operation in dynamic key updates may allow an adversary to decrypt a ciphertext without knowing the encryption key. Such a vulnerability exists when the encryption function $e_k(\cdot)$ is linear with respect to the XOR operation of its key, as shown by Eq. (4.3).

$$e_{k_1 \oplus k_2}(x) = e_{k_1}(x) \oplus e_{k_2}(x) \tag{4.3}$$

Suppose k_1 and k_2 are two key values such that $k_2 = k_1 \oplus s(t)$. The adversary maintains a short, error-free eavesdropping session and deduces $s(t)$ from the eavesdropped data frames. Suppose the same plaintext x is encrypted by both k_1 and k_2. This situation can occur in practice such as when x represents a repetitive control command. The corresponding ciphertexts c_1 and c_2 will be

$$c_1 = e_{k_1}(x)$$
$$c_2 = e_{k_2}(x) = e_{k_1 \oplus s(t)}(x). \tag{4.4}$$

Because

$$c_1 \oplus c_2 = e_{k_1 \oplus k_2}(x) = e_{s(t)}(x), \tag{4.5}$$

the adversary can decrypt x without knowing either k_1 or k_2

$$d_{s(t)}(c_1 \oplus c_2) = d_{s(t)}(e_{s(t)}(x)) = x. \tag{4.6}$$

Such a trivial vulnerability is unlikely to exist nowadays because non-linear modules are key components in the modern cipher designs [110]. It is practically infeasible to relate two ciphertexts encrypted by different keys using a modern cipher.

Suppose we cannot know the exact cipher used in a wireless LAN and have to implement dynamic key updates with the concern that the XOR operation in dynamic key updates might interfere with the cipher operations. We can replace the XOR operation in Algorithm 4 with the decryption function $d_k(\cdot)$ of the cipher and calculate k_{tmp} by

$$k_{tmp} = d_{k(t)}(s(t)). \tag{4.7}$$

The modified dynamic key update algorithm is defined as Algorithm 5. Using the modified dynamic key update algorithm, even if $s(t)$ is known to the adversary, deriving k_{tmp} is as difficult as breaking the wireless LAN cipher via ciphertext only attacks.

Algorithm 5: Modified dynamic key update

---

1 **if** *a new dynamic secret* $s(t)$ *is generated* **then**
2     $k_{tmp} \Leftarrow d_{k(t)}(s(t))$
3   **if** k_{tmp} *is not weak* **then**
4        $k(t) \Leftarrow k_{tmp}$

4.3 Proof-of-Concept Experiments

We implement the above discussed algorithms and protocols in our office wireless LAN. Our prototype implementation consists of software running on Linux computers with commercial-off-the-shelf (COTS) wireless adapters. These computers communicate with each other over an 802.11g wireless LAN. This section presents preliminary experimental results based on this prototype implementation of dynamic secrets.

The prototype programs use raw sockets [119] to control the link layer frame transmissions. In order to avoid frame retransmissions by MAC/PHY hardware, we use 802.11 broadcast frame and implement SW protocol for frame level error retransmissions. The payload size of each frame is limited to avoid link layer fragmentation. A wireless sniffer is used to verify that the frame transmissions and retransmissions behave exactly as intended by our design.

4.3.1 Prototype Design

The prototype software includes a sender program and a receiver program. We only consider the unidirectional communication scenario in which Alice sends data frames to Bob and Bob acknowledges his received data frames. Besides building two programs to handle wireless LAN link layer frames, we also tailor the standard wireless LAN frame format to suit our experiments. The frame format is specified in Fig. 4.3.

The first 6 bytes in a frame are fixed to **0xFFFFFFFFFFFF**. These 6 bytes are interpreted by the wireless LAN adapter hardware as the destination address. **0xFFFFFFFFFFFF** means the frame is a broadcast frame and should be received and processed by all wireless LAN adapters that receive it.

src_addr field contains 6 bytes as well. This field preserves its original meaning in the wireless LAN standard as the source address field. The *prt* field is set to a non-standard value to identify that the frame uses a non-standard protocol. The standard protocol stack should not process the frame.

The other fields in the customized frame format are stored in the payload field of the standard wireless LAN frame format. We put the destination address to allow the receiver to pick up the frames intended for him from all the received broadcast frames. The frame type field is used to distinguish the data frames and the acknowledgment frames. The real payload bits are placed after the sequence number field and the retransmission counter field. At the end of a frame, a 16 bits checksum is attached. We use a customized sequence number field and a customized retransmission counter

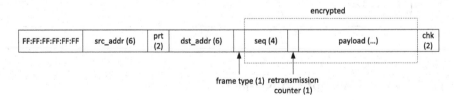

Fig. 4.3 The frame format used in the wireless LAN experiments

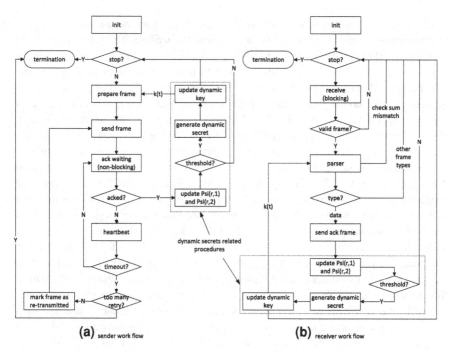

Fig. 4.4 Software work flows of the sender and the receiver programs

field instead of the corresponding fields in 802.11 frame header because we desire flexibility in control these two values through raw socket programming.

The frames are encrypted using the Blowfish cipher in the Electronic Code Book (ECB) mode [5]. The hash function used in the dynamic secrets generation, $f_H(\cdot)$, is RIPEMD-160 [45].

Figure 4.4 shows the work flows of the sender program and the receiver program. As shown in the figure, a major portion of the functional blocks and data flows are used for the frame transmissions and retransmissions. Only a small portion, shown as the red dashed rectangular areas, are related to operations for dynamic secrets generation and dynamic key updates. The integration of dynamic secrets would not significantly increase the architectural complexity of the original wireless LAN link layer communication module.

The implementation of dynamic secrets adds one data flow path into Alice's wireless LAN link layer module. Whenever a data frame is acknowledged, the sender program classifies this data frame and appends it to the corresponding frame buffer. When either the OTF buffer or the non-OTF buffer is filled up with n_{ts} data frames, dynamic secrets generation and dynamic key update will be triggered. The next sending data frame is encrypted with the updated key. For every received data frame, Bob's receiver program classifies a previously received data frame and updates the frame buffer accordingly. When the key is updated, the next received data frame will be decrypted using the updated key.

There is a small practical tweak in the receiver program because we encrypted the sequence number field as a precaution security measure. The receiver program cannot directly read the sequence number of a received data frame. When the sender program updates its key and sends a new data frame to the receiver program, the receiver program cannot use its current key to decrypt this data frame. The receiver program would not be able to know if this received frame is a new frame encrypted by an updated key or a retransmitted frame corrupted by the wireless channel. Consequently, the receiver program cannot classify the previously received data frame and trigger its dynamic secret generation and dynamic key update algorithms.

In order to make up this case, when a received data frame has the potential to trigger a key update, the receiver program would preemptively calculate the possible update key value (or two key values if $|\Psi_1| = |\Psi_2| = n_{ts} - 1$). When the next data frame cannot be decrypted by the current key, the receiver program attempts to decrypt it with the preemptive key(s). If all these keys cannot decrypt the received data frame, it will be treated as a corrupted frame and discarded by the receiver program. We pay all these complexity to reduce the number of bits that the potential adversary could manipulate. If the sequence number field is in plaintext and the hash function to generate checksum is found to be vulnerable to collision attacks, it is possible that an adversary could record a legitimate encrypted data frame and then manipulate its sequence number and checksum fields to construct a fraudulent but checksum-valid data frame. Such a fraudulent frame could be used to disturb users' communication.

In the prototype design, we allow the sender program to retransmit a frame an indefinite number of times. A practical implementation needs to have a retransmission limit. When a frame reaches this retransmission limit, the sender should save the frame in its non-volatile memory and treat the wireless LAN connection as broken. The sender will continue to retransmit this frame when the wireless LAN connection resumes. In our experimental implementation, the sender and the receiver programs store their keys and the most recently sent and received frame in hard disks. We did not implement the logic to handle the connection loss situation for the programming simplicity. A product implementation of dynamic secrets should take the cases of connection lost and resume into consideration.

4.3.2 Experiments on Computational Complexity

We firstly measure the computational cost of running dynamic secrets in wireless LAN communication. In our experiments, a sender node transmits data frames with random payloads to a receiver node. The per frame processing time for the cryptographic operations (encryption and decryption) and the dynamic secrets related operations are recorded and compared in Fig. 4.5.

We run experiments on two types of frames, short frames (40 bytes of payload) and long frames (800 bytes of payload). The experimental results indicate that the differences in payload lengths cause minimal differences in the per frame processing

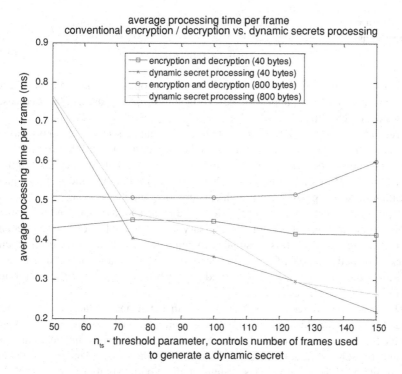

Fig. 4.5 Per frame processing time of the cryptographic operations and the dynamic secrets related operations

time for both cryptographic operations and dynamic secrets related operations. It takes slightly longer to process a long frame than a short frame.

We also vary the dynamic secrets generation threshold, n_{ts} in our experiments. The sender and the receiver would generate a dynamic secret when either their OTF buffer or non-OTF buffer is filled with n_{ts} number of packets. A change in n_{ts} does not affect the per frame processing time of the cryptographic operations. However, as n_{ts} increases, the per frame processing time for dynamic secrets related operations decreases.

The dynamic secrets related operations can be divided into two categories. One category includes operations performed for every frame such as the logic operations to classify a frame into OTF or non-OTF and the memory operations to put a frame to its corresponding buffer. The per frame processing time of this category of operations is denoted as t_α. Another category of operations consumes a fixed amount of time t_β and only runs once a dynamic secret is generated. This category of operations includes the padding operation to prepare the concatenated binary string for hashing, the hash calculations to generate a dynamic secret from the concatenated binary string of data frames, and the hard disk read and write operations for updating the key. The per frame processing time of these operations could be estimated as t_β/n_{ts}.

The per frame processing time of dynamic secrets related operations can be roughly expressed as

$$t_{ds} = t_\alpha + t_\beta / n_{ts}. \tag{4.8}$$

When n_{ts} is a small number, the second category of operations are frequently executed, and the term t_β / n_{ts} appears to be significant. As n_{ts} increases, t_β / n_{ts} decreases and t_{ds} approaches to t_α.

We compare the per frame processing time of dynamic secrets related operations and that of the cryptographic operations. When n_{ts} is small, it takes more time per frame to generate the dynamic secrets and dynamically update the key. The tipping point is $n_{ts} = 75$ as shown in Fig. 4.5. When $n_{ts} > 75$, the dynamic secrets related operations cost less processing time than encryption and decryption.

In practice, 75 frames are typically delivered in a fraction of a second in wireless LAN communications. It is often unnecessary to update the key once every second and restricting $n_{ts} \leq 75$. It is reasonable to generate a dynamic secret from several thousand of delivered data frames. In such cases, the per frame processing time of dynamic secrets related operations would be much less than that of the encryption and the decryption operations.

The experimental results suggest that, with a reasonably large value of n_{ts}, dynamic secrets would consume much less computational resources than cryptographic functions. Since the wireless LAN hardware must have sufficient computing power to execute cryptographic functions, it is likely that we could integrate dynamic secrets into existing wireless LAN systems without hardware upgrades.

4.3.3 Experiments on Adversary's Information Loss

We would also like to find out how fast a stolen key is recovered in an office wireless LAN. In this series of experiments, we use two computers to act as two legitimate users, Alice and Bob. We use a laptop based wireless sniffer to simulate the role of Eve, the adversary who constantly eavesdrops wireless communications between Alice and Bob. Our experimental settings allow Eve to stay in close proximity with Alice and Bob, a favorable condition that is often infeasible for eavesdroppers to achieve in practice.

We configure the sender program so that Alice sends a data frame to Bob every 10 ms. Eve constantly eavesdrops the wireless signals. We measure the time that Eve encounters the first frame loss and the 10th frame loss for data frames. We run this test multiple times with different payload lengths. Figure 4.6 shows the mean time we recorded in the tests and the standard deviations.

As shown in Fig. 4.6, Eve encounters frame losses shortly after Alice begins sending frames. The frames with larger payloads are more likely to be lost. For all payload lengths, Eve's first frame loss occurs on average within a second. The average time for Eve to lose 10 frames is less than 5 s.

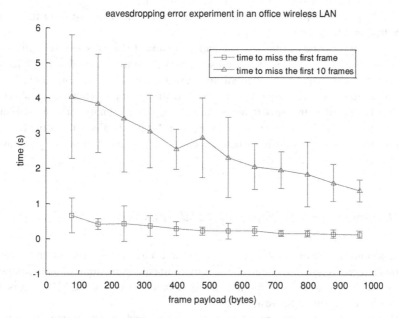

Fig. 4.6 Tests on eavesdropping packet losses in an indoor office wireless LAN

The experiments also indicate that even when Eve eavesdrops from a location very close to Alice, which provides a high SNR for Eve's receiver, frame losses are non-negligible. This is because the indoor wireless environment is interference dominating [101]. Eve may try to mitigate the interferences by building an eavesdropping device using advanced technology such as Multiple Input Multiple Output (MIMO) antenna array [105]. However, such a dedicated device is not only expensive, but also exposes Eve to much higher risk of being noticed and caught.

Reference [111] presents a wireless LAN eavesdropping experiment with its settings being artificially constructed in favor for the eavesdropper. In the experiment, four eavesdropping devices are placed within one meter radius of the eavesdropping target in an anechoic chamber. These four eavesdropping devices work in a colluding way. Only when a frame is lost by all four devices, the frame is deemed as lost by the eavesdropper. Moreover, the anechoic chamber minimizes signal interferences and environment noises for eavesdropping. Although the experiment settings are extremely favorable, the eavesdropper still observe non-negligible packet loss within 20 s.

The experiment results further suggest the possibility of establishing a shared secret key between Alice and Bob without any prior shared secret information, i.e. the possibility of key bootstrapping in wireless LAN.

Before communication, Alice and Bob publicly agrees on an initial value for key bootstrapping, such as $k(0) = 0$. Alice then begins sending data frames to Bob with random payloads from time 0. Eve would not learn any sensitive information by eavesdropping these frames.

After a short period of time t, Alice and Bob know that $k(t)$ is a secret with over-whelming probability. Then they can use $k(t)$ as the initial key value for dynamic key and start exchanging confidential information in their secure communications. According to our experimental results, the bootstrapping period can be as short as several seconds in an indoor wireless LAN. If Alice keeps sending frames with random payloads to Bob for more than 30 s, it would be practically impossible for Eve to know the initial dynamic key when Alice and Bob begins their secure communication.

4.3.4 Experiments on Environmental Randomness

An important feature of dynamic secrets is to harvest true randomness from the communication environment and convert the environmental randomness to randomness of key values. We would like to know that in a practical wireless LAN, what is the rate of true randomness harvest.

We configure the sender program so that Alice sends a data frame once every second to Bob. The threshold parameter n_{ts} is set to 60. The experiment runs for a 24-h period. Figure 4.7 summarizes the experimental results for a 24-h period.

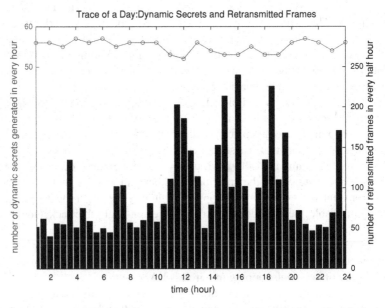

Fig. 4.7 Dynamic secrets generation and dynamic key updates, logged traces of a day ($n_{ts} = 60$)

In the figure, the number of retransmitted frames reflects the noise and interference levels in the wireless communication environment, e.g. level of environmental randomness. We can calculate the frame loss rate from the number of retransmitted frames. The frame loss rate averages around 2.5 % and varies throughout the day.

Recall the analytic results in Sect. 3.3. We can calculate the amount of true randomness harvested for an arbitrary number of buffered frames. By looking back to the logged trace of frame transmissions, we find that approximately 14500 bits of true randomness have been harvested in the experiment, even with such a low rate of one frame per second.

The generation of dynamic secrets remains stable throughout the tests although the randomness level changed substantially in the 24 h experiment. 50–60 dynamic secrets are generated every hour. The key is updated roughly every minute.

4.4 Dynamic Secrets in Other Layers of Wireless Communications

In the remainder of this chapter we discuss how to extend the usage of dynamic secrets from the link layer to other layers of wireless communications. Different layers and different wireless communication scenarios may impose different requirements on the dynamic secrets implementation.

4.4.1 Throughput Efficient Implementation in Network Layer

The idea of dynamic secrets was presented in a packet communication model in Chap. 3. Therefore the network layer adoption of dynamic secrets is natural. The software architecture of dynamic secrets in the network layer is similar to that in the wireless LAN link layer.

It is possible to implement the SW protocol based algorithms in the network layer and to just mimic the wireless LAN link layer implementation. However, SW may substantially limit the network layer communication efficiency.

The round trip time (RTT) of a network layer packet is often much longer than that of a link layer frame. The typical wireless LAN link layer RTT is in the order of tens of microseconds while the network layer RTT often ranges from tens to hundreds of milliseconds. If Alice stops and waits for a packet acknowledgment before sending the next packet, Bob may only receive several packets per second when the wireless channel is noisy and there are significant packet losses.

Here we propose a throughput-friendly implementation of dynamic secrets in network layer. The sender, Alice, maintains a sufficiently large packet buffer Ψ_s. The receiver, Bob, maintains a packet buffer Ψ_r that can contain n_{ts} number of packets. Before Alice and Bob communicate, they share a symmetric key k and set $\Psi_s = \Psi_r = \phi$.

Algorithm 6: Dynamic secrets in network layer: sender

1 **foreach** *packet m_i to send* **do**
2   transmit m_i;
3   add m_i to Ψ_s;
4   **if** *a key update notification packet m_{ku} is received from Bob* **then**
5   select and rearrange packets according to information in m_{ku};
6   remove the selected packets from Ψ_s;
7   generate dynamic secret s;
8           $k \Leftarrow k \oplus s$;
9   retransmit all packets left in Ψ_s;

Algorithm 7: Dynamic secrets in network layer: receiver

1 **foreach** *received packet m_i from Alice* **do**
2   add m_i to Ψ_r;
3   **if** $|\Psi_r| == n_{ts}$ **then**
4   Rearrange packets in Ψ_r according to their retransmission flags;
5   Hash the rearranged packets to generate dynamic secret s;
6           $k \Leftarrow k \oplus s$;
7   Form a key update notification packet m_{ku} that contains
8   sequence numbers and retransmission flags of packets in Ψ_r;
9           $\Psi_r \Leftarrow \phi$;
10   **repeat**
11   transmit m_{ku};
12   discard all received packets that is not secured by the updated k;
13   **until** *a new packet secured by the updated key k is received*;

In Algorithms 6 and 7, Alice continuously transmits data packets to Bob until Bob notifies her to update the key. Bob continuously adds received data packets to his buffer, Ψ_r, until his buffer is filled up with n_{ts} packets. Bob generates a dynamic secret and updates his key. Then he sends a notification packet, m_{ku}, to Alice that contains the sequence numbers of the n_{ts} packets in his buffer. Bob repeatedly sends m_{ku} and discards any incoming packet encrypted by the previous key, until a data packet encrypted by the update key is received. At that time, he knows that Alice has updated her key as well.

Alice continuously sends data packets to Bob using the same key to encrypt them until she receives the key update notification packet m_{ku} from Bob. When Alice receives m_{ku}, she updates her key using the sequence number information contained in m_{ku} and removes all of the packets that have been used in the key update. She then retransmits all the packets that are left in Ψ_s and continues to send new packets.

Because Alice does not need to wait for a particular acknowledgment packet before sending more packets, she can send batches of data packets and efficiently utilize the bandwidth. Some bandwidth is wasted after Bob's key update and before Alice's key update. It is possible to design more complex and more efficient algorithms to implement dynamic secrets in the network layer.

It would be interesting to extend other reliable transmission protocols such as the Go-Back-N protocol and the Selective Repeat protocol to adopt the idea of dynamic secrets. However, this task is not straightforward because dynamic secrets not only demand reliable packet delivery but also request both Alice and Bob to be aware of whether a packet is delivered through retransmissions or not. We will leave it as a future research topic.

4.4.2 Application Layer Dynamic Secrets, Leasing and Proxying

Users can implement dynamic secrets at the application layer and obtain the security benefits brought by dynamic secrets. The transmission units in the application layer are messages. Dynamic secrets work flow for the packets and the link layer frames can be used for the application layer messages as well.

When implementing an application layer software with dynamic secrets, we do not need to consider the compatibility issues that must be taken care of in the lower layer implementations where transmission protocols and cryptographic procedures already exist. It is quite flexible to write a secure communication application to make use of dynamic secrets. On the other hand, the programming could be complicated because it would handle the naming and addressing issues, support reliable message transfers, and perform cryptographic operations.

Because of the flexibility of the application layer implementation, here we do not discuss the implementation details such as algorithms, protocols, and program workflows. Instead, we present three possible architectures for an application layer implementation of dynamic secrets, as shown in Fig. 4.8.

The first architecture provides the greatest flexibility because each communication application can implement dynamic secrets according to its own requirements and preferences. This flexibility comes at the cost of programming complexity. Each application has to implement and maintain a dynamic secrets framework and a traditional secure message transfer framework using the secret key. Another concern of this architecture is that all applications have to maintain a particular level of communications. If an application rarely exchange messages between Alice and Bob, it will not gain much security benefits from dynamic secrets.

In the second architecture, Alice and Bob each maintains a leaser application which constantly transfer random messages to each other and use dynamic secrets to update the key. When Alice's application wants to secure communicate with Bob's application, they lease a short term secret key from the leaser application. The leased key is a temporary key derived from the leaser's dynamic key at the time of the lease. In this architecture, the communication applications can focus on the traditional secure

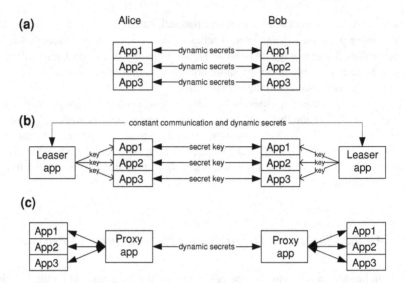

Fig. 4.8 Three application layer implementation architectures: **a** application-to-application, **b** dynamic key leasing, **c** dynamic secrets proxy

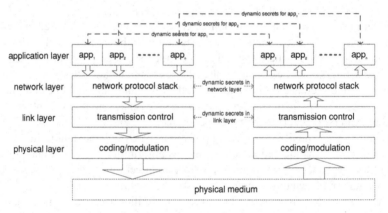

Fig. 4.9 Using dynamic secrets in multiple layers to provide strong protection for the wireless communication system

message transfer process. Key management is outsourced to the leaser application. Also the applications can maintain a low level of communication activity but still enjoy a high level of dynamic secrets protection, at the cost of extra bandwidth consumed by the leaser application's communication of random messages.

The third architecture calls for an application layer communication proxy. The proxy application implements both the traditional secret key based secure message transfer framework and the dynamic secrets framework. The other applications send and receive messages through the proxy application. This architecture has the minimal programming complexity among all three architectures. Moreover, it does not consume any additional bandwidth. Dynamic secrets are generated from the aggregated communication traffic of all applications.

It is possible to integrate different layers of dynamic secrets implementations into one secure wireless communication system. Figure 4.9 illustrates a system design that simultaneously uses multiple layers of dynamic secrets. Because each sender and receiver pair at different layer manage their dynamic secrets and dynamic key independently, such an integration would not cause an excessive increase in system complexity.

Chapter 5
Dynamic Key Management in a Smart Grid

This chapter extends the discussion of dynamic secrets from the secure communication between two wireless users to the cryptographic key management in a large scale networked environment. Specifically, this chapter chooses smart grid as an application scenario to explore the practicality of the key management scheme based on dynamic secrets. The deployment cost, the management complexity, and the scalability issues are addressed.

Smart grid is a vital integration of the traditional power grid and a communication network that enables real time information sharing and control across the power grid. The communication capability among grid devices enables higher power utilization efficiency for a smart grid than its traditional counterpart.

Smart grid is an emerging infrastructure. There are a great variety of proposals for the underlying communication technology used in smart grid. A main stream choice is to enable wireless communication capability on smart meters so that they can make use of widely deployed mobile wireless infrastructure or wireless LAN access to report their measurement data to the grid control center and receive commands from it. Power line communication (PLC) is an attractive alternative to wireless communication. However, PLC has an inherent limitation that affects its deployment. PLC requires relay devices to pass PLC signals through a power transformer. This requirement substantially increases the deployment cost and maintenance complexity of PLC in smart grid. In this chapter we make a primitive proposal for certain aspects of smart grid security assuming wide use of wireless communication technology in smart grid devices.

There are many difficult problems need to be solves in order to architect a secure communication infrastructure that can protect a smart grid from cyberphysical security threats [61]. One major challenge is to design an efficient, scalable cryptographic key management scheme (KMS) for smart grid communication networks. A smart grid may contain millions of networked nodes. A KMS in such a scale is extremely expensive, if not infeasible, according to our Internet security experience. These nodes can be distantly scattered and have severely limited accessibility. For example, some electrical generators may be located in a high latitude

S. Xiao et al., *Dynamic Secrets in Communication Security*,
DOI: 10.1007/978-1-4614-7831-7_5,
© Springer Science+Business Media New York 2014

region where snow blocks the access road in winter months. These nodes would be difficult to be secured if on-site configuration is required.

The smart grid network environment is less favorable for security than a typical computer network. A KMS designed for a traditional computer network will encounter performance hurdles or even be infeasible when directly applied to a smart grid communication network [95]. On the other side, smart grid demands for a more secure KMS than computer networks because the loss of information security in smart grid may lead to substantial physical world damages.

This chapter presents dynamic key management scheme (DKMS) as a lightweight,scalable key management solution for the smart grid communication network. The automatic stolen key recovery feature of dynamic secrets makes DKMS more robust than conventional smart grid KMS schemes.

Section 5.1 briefly overviews a smart grid communication architecture and discusses challenges in designing a smart grid cryptographic key management scheme. Section 5.2 presents the basic building block of the dynamic key management scheme, the dynamic key update protocol used between two grid nodes. We use communications between a smart meter and its metering report collector as the example to demonstrate the dynamic key update protocol. We also analyze the computation and bandwidth costs of the dynamic key update operations. Section 5.3 provides a grid level overview of the dynamic key management scheme and discusses the operation details of node installation, removal, and the automatic process used between two remote nodes to establish a secure end-to-end communication. Section 5.4 further discusses the scalability of the dynamic key management scheme and the crisis responses when the grid communication infrastructure is attacked.

5.1 Smart Grid and Cryptographic Key Management

We model the structure and essential elements of a smart grid as a network shown in Fig. 5.1. There are three elements in our smart grid model: power generators, power transmission network, and the utility meters. Electricity flows from the generators through the power transmission network to the meters.

In a traditional power grid, the grid operators need to oversee load balancing and provision the electricity generation to satisfy the demands of the end users [4]. The information available to the grid operators are often coarse-grained and delayed, such as the meter reports manually collected at the end of each month. It is difficult for the grid operator to optimize the energy efficiency based on users' real time demands. Moreover, when the electrical load suddenly changes or the key connection in the power transmission network fails, the grid may not react quickly to compensate the mismatch between the power generation and the electrical load. The unbalanced grid may experience cascading failures and result in a large area blackout [6, 62].

In a smart grid, the power generator, the meter, and the control devices in power transmission network are interconnected in a communication network. These networked devices can act "smartly" according to real time information feedback from

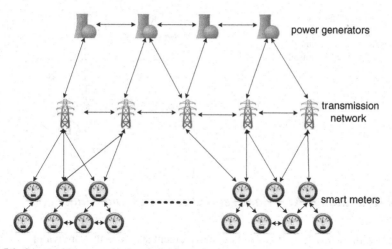

Fig. 5.1 Smart grid communication network

other devices. The real time information sharing provide the opportunity for a smart grid to significantly improve energy efficiency over a traditional power grid. Moreover, a smart grid could react agilely to device failures and prevent a local failure from propagating and causing a grid level disaster.

The communication among smart grid nodes are bi-directional and persistent. For example, a smart meter needs to periodically report its utility consumption to upstream power distribution devices. A grid control unit may also periodically check this smart meter to ensure it works properly and can correctly react to control commands. Among smart meters in local proximity, they may exchange information and act in a cooperative manner to optimize local electricity distribution. A power generator needs to constantly exchange information with the power transmission network to adjust its electricity output to match the users' power consumptions. The devices in power transmission network are constantly sending and receiving information with each other to optimize the overall power delivery efficiency.

The physic links between smart grid devices are often wireless because the devices are likely to be geographically scattered. It is expensive to establish a dedicated, wired communication network infrastructure to support the information exchange in the power grid. There are other types of physical links such as power line communications, telephone lines, and cables. In practice, these types of physic links would have non-negligible information loss in communications [32]. The discussion based on wireless links could be applicable to these types of links as well.

We can summarize our smart grid communication network model as the follows:

- The network could have millions of nodes and an on-site configuration for every node is hardly feasible.
- The communication link between nodes is *noisy*.
- The communication traffic between nodes is *bidirectional* and *persistent*.

Fig. 5.2 The structures
of three key management
schemes: **a** key server based,
b point-to-point, **c** PKI based

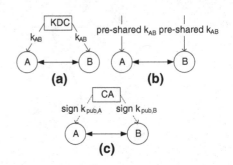

5.1.1 Traditional Key Management Schemes and Challenges

Typically, there are three types of key management schemes that are widely used in network security. They are Key server based KMS, Point-to-Point KMS, and Public Key Infrastructure (PKI) based KMS. The structures of these traditional KMSs are illustrated in Fig. 5.2.

A and B represent two connected nodes in a communication network. In the key server based KMS, node A and node B request a key k_{AB} from a trusted key distribution center (KDC) before they communicate with each other. node A and node B will request a new key for each communication session. The secure links KDC-A and KDC-B must be constantly available because node A and node B may demand a secure communication anytime.

Deploying a key server based KMS in a smart grid would be difficult. Every smart grid node would require a constant available, secure link to request key from KDC. Such an infrastructure would be prohibitively expensive for millions of geographically scattered smart grid nodes. The key server based KMS fits the requirements of a small, localized communication network, such as a university campus [9], but is hardly scalable to a smart grid. There are key server based KMS designs that aim to improve scalability [3]. The extensions involve public key infrastructure and inherit the limitations of PKI based KMS. For example, it is hard to solve the public key certificate compatibility problem among different organizations in a smart grid.

Point-to-pint KMS structure assumes that node A and node B have a pre-shared long term symmetric key k_{AB}. Before each secure communication session, node A and node B use k_{AB} to securely exchange a short term session key. Compared to key server based KMS, point-to-point KMS does not require the constant availability of a supportive infrastructure.

A notable problem for point-to-point KMS to be deployed in smart grid is the difficulty of key distribution and destruction. When a smart meter is installed in the grid, long term symmetric keys need to be installed on this smart meter and all the devices that this smart meter may communicate with. Some of these devices might locate far away and hard to access. Moreover, it is hard to provision which devices would have communication with the newly installed smart meter. In the future, if the smart grid wants to change its communication topology, the grid operator must send service personnels to every grid node to reinstall the long term symmetric keys.

Even if the installation of keys is feasible, removing this smart meter from the smart grid will cause a similar problem. In order to avoid an adversary using the keys stored in the removed smart meter to send fraudulent messages to the grid, all the long term symmetric keys stored on other devices and used to secure communicate with the removed smart meter have to be destructed. The key destruction is as costly as the key distribution.

PKI based KMS uses a certificate authority (CA) to sign and authorize node A's and node B's public key digital certificates. A signed digital certificate is installed in each node when it is manufactured. Using public key cryptography, node A and node B can authenticate each other by verifying their digital certificates and then exchange short term symmetric keys to secure their communications.

If node A stores the public key of the CA that signs on node B's digital certificate, node A can verify node B's identity without consulting any other devices. The same condition applies to node B. If all nodes in a network share the same CA, the secure communications between any two nodes would not require the presence of CA.

PKI based KMS avoid some drawbacks of key server based KMS and point-to-point KMS. On one side, PKI based KMS does not require constant secure connection between a node and CA. Two nodes could establish a secure connection by themselves. On the other side, the installation of a node does not involve configuration of keys on other nodes. PKI based KMS has its own problems when applied to a smart grid. Three notable problems are addressed in Ref. [118].

The first problem is called *trust roots*. If node A does not have the public key of the CA that signs on node B's digital certificate, node A would not be able to verify the authenticity of node B's certificate. Consequently, node A could not have secure communication with node B. In the textbook version PKI, this problem does not exist because it assumes a universal CA is used for everybody. In a smart grid, devices are produced by hundreds, if not thousands, of manufacturers from all over the world. It is impractical to coordinate all the manufacturers and pre install public keys from all CAs to the smart grid devices.

The second problem is called *trust paths*, which is also studied in Ref. [1]. A node's digital certificate may be signed by a chain of intermediate nodes before the trust path reaches a top level CA. Therefore, verifying this node's digital certificate involves verifications of all the intermediate nodes' certificates. This chained verification process may result in long delays in communication. This effect is detrimental to the real time requirement of the smart grid communications.

The third problem is called *revocation*. When a node is to be removed from the smart grid. The digital certificate of this node must be revoked. The revoke process requires a broadcast announcement to all the other nodes in the smart grid. Such an operation is extremely expensive and impractical in a large scale smart grid.

In Ref. [16], Baumeister et al. presents an overview for the requirements for a smart grid key management scheme and the capability of PKI based KMS. This study concludes that current PKI based KMS used in computer networks cannot satisfy the security and practicality requirements of smart grid.

5.2 Dynamic Key, a Node-to-Node Autonomy

Dynamic secrets can be used to create a decentralized, lightweight key management solution for smart grid, named as dynamic key management scheme (DKMS). The basic idea of DKMS is to use a protocol similar to wireless key bootstrapping in Chap. 4 to generate a dynamic key between two smart grid nodes and then maintain dynamic key updates using algorithms introduced in Chap. 3.

The bootstrapping of dynamic key help removing the requirement of infrastructural support such as KDC or CA. We will show that the key dynamics can be used for effort-less key destruction as well. DKMS avoid the trust root and trust path problems of PKI based KMS because dynamic keys are generated after devices installed into the smart grid instead of pre-set at manufacturing time.

This section focuses on the dynamic key update protocol between two smart grid nodes, more specifically, between a smart meter and its metering report collecting device. This protocol is used for key updates as well as key generation. Both protocol design and security characteristics are included. The dynamic key update protocol presented in this section is the fundamental building block of DKMS.

5.2.1 Key Updates Between Meter and Collector

We assume that the smart meter periodically sends its metering report to the collector device. The communication channel is noisy and may corrupt the messages transferred through the channel. The noise effect is ubiquitous and even an adversary equipped with advanced eavesdropping technology cannot avoid information loss.

Dynamic key update protocol is defined in Protocols 1 and 2. Ψ_C is the message buffer maintained at the report collector device. Ψ_M is the message buffer in the smart meter. m_i represents a metering report message with sequence number i.

Protocol 1 Smart meter's key update protocol

---
if metering report m_i is to be sent **then**
    append all unacknowledged reports with m_i
    encrypt the reports with key k
    send encrypted reports along with the sequence numbers of each report in clear text
    add m_i to Ψ_M
end if
if an acknowledgment is received **then**
    calculate $s = f_h(\Psi_M)$
    update $k \Leftarrow f_h(k||s)$
    empty Ψ_M
end if

Protocol 2 Report collector's key update protocol

---

if one or more metering reports is received **then**
  send corresponding acknowledgment(s)
  if all received reports are not acknowledged before **then**
    calculate $s = f_h(\Psi_C)$
    update $k \Leftarrow f_h(k\|s)$
    empty Ψ_C
  end if
  decrypt the received reports as $\{m\}$
  add the unacknowledged reports in $\{m\}$ to Ψ_C
end if

Figure 5.3 illustrates an example key update process. In the following discussion, we use $E_k(x)$ to denote the encryption of plain text x using key k. When the smart meter sends a message m_i to the collector device, it sends the sequence number in plain text and the encrypted message as a tuple $\{i, E_k(m_i)\}$. The smart meter uses

Fig. 5.3 An example key update process between the smart meter and its metering report collecting device

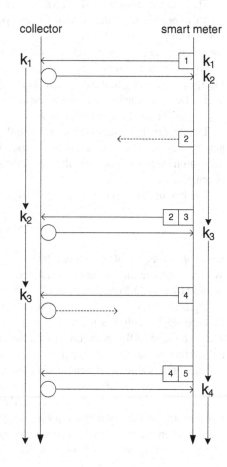

message integrity check to ensure that adversary cannot modify the sequence number and resend the message.

The smart meter sends $\{1, E_{k_1}(m_1)\}$ to the collector, and then add m_1 to Ψ_M. The collector acknowledges the delivery of m_1, decrypt m_1 using k_1, and add m_1 to Ψ_C. The acknowledgment successfully arrives the smart meter. It triggers the smart meter to update its key to k_2. After the key update, the smart meter empties Ψ_M.

The smart meter sends $\{2, E_{k_2}(m_2)\}$ to the collector, and then add m_2 to Ψ_M. The transmission of m_2 is disrupted by the noise in the communication link. The collector does not receive m_2 and would not acknowledge.

Upon the time to report m_3 to the collector, the smart meter finds out that m_2 has not been acknowledged. Therefore the smart meter sends $\{2, 3, E_{k_2}(m_2||m_3)\}$ and add m_3 to Ψ_M. The collector then acknowledges m_2 and m_3 and updates its key to k_2 using $\Psi_C = \{m_1\}$. After the key update, the collector sets $\Psi_C = \phi$ and uses k_2 to decrypt m_3 and m_2. m_2 and m_3 are then added into Ψ_C. The acknowledgment of m_2 and m_3 are received by the smart meter. It update its key to k_3 using $\Psi_M = \{m_2||m_3\}$. After the key update, $\Psi_M = \phi$.

The smart meter sends $\{4, E_{k_3}(m_4)\}$ to the collector, and then add m_4 to Ψ_M. The arrival of m_4 triggered the collector's key update. After the key update, the collector's key becomes k_3 and $\Psi_C = \{m_4\}$. The collector sends out the acknowledgment for m_4, but the smart meter fails to receive it.

From the smart meter's point of view, m_4 is unacknowledged. Therefore it sends $\{4, 5, E_{k_3}(m_4||m_5)\}$ to the collector. The collector receives both m_4 and m_5. Because m_4 has been acknowledged before, the key update is not triggered. The collector adds m_5 to Ψ_C. At this moment, $\Psi_M = \Psi_C = \{m_4||m_5\}$.

The acknowledgment for both m_4 and m_5 arrives at the smart meter and triggers its key update. The smart meter uses k_4 to encrypt the next status report. As the communication goes on, the collector and the smart meter updates their keys alternatively.

The key update from k_i to k_{i+1} can be expressed as

$$k_{i+1} = f_h(k_i || f_h(\Psi)), \tag{5.1}$$

where $\Psi = \Psi_C = \Psi_M$ denotes the message buffer at the moment of key update. We can compare the key updates defined by Eq. (5.1) with the dynamic key updates defined by Eqs. (3.1) and (3.2) in Chap. 3.

Equations (3.1) and (3.2) update a key by XORing with a dynamic secret. Equation (5.1) computes the hash function twice for key updates. There are two reasons to replace XOR operation with hashing. Firstly, the metering reports would have a fixed format and the data are likely to be correlated, i.e. the power usage in the past fifteen minutes and the fifteen minute starting from current time. Although the use of sequence number ensures that identical measurement data would yield different hash values and therefore the hash values would not cancel each other in XOR key updates, it is possible that the outcomes of bitwise-XOR key update exhibit some exploitable pattern when the smart meter keeps sending identical measurement data. We would like to be conservative and use hash function to further "mix" the hash value of a metering report and the key to be updated. Secondly, dynamic key update algorithm in Chap. 3 is proposed for packet communication scenarios. The packet sending and receiving rate might be high and key updates could be frequent. In these scenarios, XOR operation is preferred because it costs less computation resources than hashing. In the smart meter case, the typical time gap between two metering reports are 15 min [74]. The smart meter and the collector device would have sufficient time to perform two hash operations for a key update.

5.2.2 Security Properties of Dynamic Key Updates

The dynamic key update protocol between the smart meter and the collector device inherits all three security features as those provided by dynamic key updates in noisy packet communications. Recall the discussion in Chap. 3. The three security features are:

1. true randomness in key values;
2. automatic recovery for stolen key;
3. intrinsic detection for impersonation attacks.

A metering report may be corrupted in its transmission and then attached with the next metering report for a retransmission. Depending on the transmission process of metering reports, Ψ may contain one or more metering reports to be hashed. No one can predict the content in Ψ because physical randomness in the communication channel affects the metering report transmission process and consequently, the content in Ψ. The content of Ψ contains true randomness. Through the key updates, the true randomness accumulates in the key and prevents the adversary from cracking the key.

According to studies [80, 97, 106, 114], smart grid communication channels have non-negligible packet loss rate even at ideal testing conditions. To maintain an error-free eavesdropping for a long period of time is practically impossible. Moreover, the smart grid devices may be installed in hardly accessible places and aggravates the cost of eavesdropping.

When adversary fails to eavesdrop every bit of metering reports, he will surely fail to keep tracking the dynamic key updates. As a result, the stolen key in smart grid would be automatically recovered. The key recovery process in smart grid is similar to the key recovery in noisy packet communications, as discussed in Chap. 3. It is noteworthy that in the smart meter and collector device case, the stolen key recovery could take days against a patiently eavesdropping adversary because the smart meter communicate with the collector device only once every 15 min.

Suppose an adversarial user in a smart grid wants to reduce his monthly electricity bill. He will need the cryptographic key in his smart meter to submit fake metering reports. The adversary may not be able to directly read the key because smart meter usually equips with temper-proofing key storage hardware. However, he may use side channel attack to circumvent the hardware protection or use social engineering attack to obtain the key. He then use a wireless enabled computer to construct and send out a faked metering report to the collector using the stolen key.

If the key is a traditional symmetric or asymmetric key. the adversary would have been successful on his malicious plan. The report collecting device will treat the legitimate report sent by the smart meter as a duplicate transmission and discard it. It is difficult to notice such a forgery.

If the key is a dynamic key, the resulted key update in the collector device would block the smart meter from submitting metering reports in the future. Then the smart meter would alarm and signal the abnormal situation. Any further check would surely reveal the existence of the impersonation attack. The detection to impersonation attacks is intrinsic and accurate. Because the smart meter persistently submit metering reports days and nights. Impersonation attacks would always be detected.

5.2.3   Computation, Storage, and Bandwidth Cost

The computation of a key update costs two hash computations when two smart grid nodes maintain a low rate communication and update their key infrequently, such as the above meter-collector example. The above key update protocol may also be applied to the secure communications between wireless sensors and their monitors in a smart grid. In high speed communication scenarios such as the data exchange between two grid control centers, we may apply more efficient key update protocols such as the one defined in Chap. 4.

The storage requirement is case-specific. For the above key update protocol used between a smart meter and a report collector, we may estimate the maximum buffer size the meter and the collector needs to have. Let L denote the size of a metering report. T_r denotes the time between two metering reports. A typical value for T_r is 15 min. T_c denotes the maximum period of time allowed for a smart meter to be "absent" from its report collector. If a collector has not seen any metering report from a smart meter for more than T_c time, the collector would think the smart meter as having been removed from the smart grid and destroy the key used to securely communicate with this smart meter. The message buffer size should not be smaller than LT_c/T_r because it is possible for a smart meter to fail $T_c/T_r - 1$ times of report transmissions in a row and finally send a bulk of T_c/T_r number of metering reports together through a bad communication link.

A metering report only has several kilo bytes data as suggested in Ref. [15]. Suppose a collector can tolerate a smart meter three days without report at most. The lower bound of the message buffer size is roughly three mega bytes which is easily implementable on a smart meter with today's storage technology. A collector device may need to collect metering reports from many smart meters. However, the collector may use its message buffer in a multiplexing way. It would be over-conservative to require the collector device to maintain a full size, independent message buffer for each smart meter. We may consult the knowledge of Internet router design to estimate a reasonable buffer size for the report collector device.

In the above key update example, the smart meter and the collector device synchronize their dynamic key updates without any extra communication. For other types of node communications, they may need extra communication to synchronize their key updates. The cost of key update synchronization is controllable because the two nodes in communication can lower their key update frequency to limit the resources consumption.

5.3 Node Installation, Node Removal, and Trust Propagation

This section discusses the key management operations associated with installing and removing a smart grid node. We demonstrate a low cost approach for key management in node installations and removals. Furthermore, we propose a key distribution method namely trust propagation that allows two nodes to establish an end-to-end secure communication channel with the assistance from other nodes.

5.3.1 Node Installation and Removal

When a node is newly installed, it is necessary to initialize a dynamic key between this node and another node that already exists in the smart grid. We name the newly installed node as node A and the existing node as node B. The choice of node B could be arbitrary as long as the communications between node A and node B are not relayed by a third node in the smart grid.

There are two ways to initialize the key between node A and node B, as shown in Fig. 5.4. Firstly, an initial key value $k(0)$ can be manually setup when node A is being physically installed. The installation personnels can manually input a $k(0)$ to both nodes. $k(0)$ does not have to be truly random or perfectly secret. Later, node A and node B will exchange random messages for a period of time and update their

Fig. 5.4 Two methods to initialize a dynamic key when a node is installed. **a** The installation personnels negotiate $k(0)$ using their cell phones. **b** Key bootstrap from a public default $k(0)$

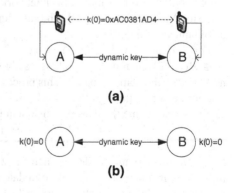

dynamic key many times. In this way, the key secrecy is improved. Moreover, the key value would not be deducible from $k(0)$. An adversary cannot obtain the key by bribing the installation personnel. Secondly, node A and node B can bootstrap their key from a publicly known default value, e.g. $k(0) = 0$. In order to obtain a sufficiently strong key for the secure communication, node A and node B need to exchange random messages for a longer period of time than that in the first method.

The key destruction associated with node removal is very simple in the dynamic key management scheme. A node maintains a watchdog timer for each dynamic key it manages. Whenever a key is updated, its timer will be reset. If a timer runs out, the node discard the corresponding key. Therefore, when a node is removed from a smart grid, it is not necessary to delete keys from other nodes. Other nodes will gradually "forget" the removed node and discard the associated keys.

The time limit of a watchdog timer should be set according to the communication scenarios. The report collector can set a watchdog timer of 72 h to monitor the key between itself and a smart meter. When a key is used for the secure communication between a power generator and a voltage regulator, the watchdog timer limit may be as low as several seconds because the power generator and the voltage regulator would exchange information very frequently. Moreover, if a power generator does not hear from its voltage regulator for several seconds, its power output might already be out of control and actions must be taken for crisis response.

5.3.2 Trust Propagation and End-to-End Secure Communication

When node A is newly installed, it only has one dynamic key with node B. We say node A and node B trust each other when they maintain a dynamic key for their secure communications. Node A may want to securely communicate with more nodes and establish more mutually trust relationships in the smart grid.

Let $T(A)$ denote the set of nodes that node A trusts. We have $B \in T(A)$. Because the trust is a mutual relationship, $B \in T(A) \Rightarrow A \in T(B)$.

Let $P(X, Y)$ represent a set of nodes on a communication route between node X and node Y without loop. For a network topology shown in Fig. 5.5, a possible value for $P(A, D)$ is $\{A, C, D\}$. The solid lines in the figure represent direct physical communication links.

For a node X that $X \in T(B)$ and $X \notin T(A)$, if there exists a $P(A, X)$ such that $B \notin P(A, X)$, then it is possible to establish a dynamic key between node A and node X with the help of node B. This process is like a person A gets acquainted with another person X through the introduction of person B who knows both A and X. Therefore, we name this key establishment process as *trust propagation*.

As shown in Fig. 5.5, node A and node C can establish a dynamic key k_{AC} through the help of node B and using Protocol 3. Node B distributes a random, initial key $k_{AC}(0)$ to node A and node C. Then these two nodes transmit random messages through their direct physical link to update their key k_{AC}. Although node B knows $k_{AC}(0)$, it would not be able to deduce k_{AC} after a while. Node A's trust set is then

Fig. 5.5 Trust propagation in
a smart grid using dynamic
key management scheme

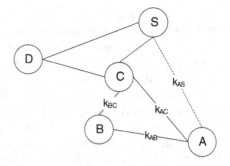

Protocol 3 Trust propagation for nodes X, Y, and Z

---

node X query node Y for $T(Y)$
node X find node Z such that $Z \in T(Y)$ and $Z \notin T(X)$
node X find a communication route $P(X, Z)$ such that $Y \notin P(X, Z)$
node X request node Y to help with trust propagation
node Y randomly generate a initial key $k_{XZ}(0)$ and securely send it to node X and node Z
node X and node Z exchange random messages through $P(X, Z)$ and update k_{XZ} from $k_{XZ}(0)$
after a period of time, node X and node Z use k_{XZ} for their secure communications

---

expanded, from $T(A) = \{B\}$ to $T(A) = \{B, C\}$. If node D is trusted by node B, i.e.
$D \in T(B)$, node A can further propagate trust to node D through the communication
route $P(A, D) = \{A, C, D\}$.

However, if $D \notin T(B)$, node A will not be able to propagate trust to node D
because all routes between node A and node D involves node C. In this case, node
A can request help from the grid operator to manually setup a dynamic key between
node A and a node S that $D \in T(S)$. Node S randomly generates an initial key $k_{AD}(0)$
and distribute $k_{AD}(0)$ to node A and node D. Although node S sends $k_{AD}(0)$ to node
A through the relay of node C, node C does not know the value of $k_{AD}(0)$ because
the messages exchanged between node S and node A are end-to-end encrypted using
the key k_{AS}.

5.4 Scalability and Crisis Response

Dynamic key management scheme is completely decentralized that every node
autonomously manages the dynamic keys used for its secure communications. Moreover, dynamic keys are mutually independent to each other. If each node has sufficient
resources to manage its keys, dynamic key management scheme does not have an
upper limit for the total number of nodes that the scheme can support.

As a comparison, the scalability of key server based KMS is limited by the total
number of nodes that the KDC can handle. Point-to-point KMS requires manual key
configuration for every node. Therefore it is inherently not scalable. PKI based KMS

in theory has great scalability. However, if a signed digital certificate is to be revoked, CA has to broadcast a certificate revoke list (CRL) to every node in a smart grid. This operation becomes infeasible when the number of nodes is large.

Dynamic key management scheme has better crisis response performance than traditional KMSs. Firstly, dynamic key management scheme is more resistant to the crisis associated with the cryptographic keys. The decentralized nature of DKMS eliminates the risk of single point of failure. Each node only stores the keys used by itself. The key theft to a node only affects the communication security between this victim node and its communication peers. In key server based KMS and PKI based KMS, if the KDC or the top tier CA is compromised by the adversary, communication security of entire smart grid is affected.

Secondly, dynamic keys can automatically recover from key theft and traditional key cannot. Suppose the adversary intrude a grid control center and obtain all the keys that are used to encrypt and send the controlling commands to the devices in a smart grid. When a traditional KMS is used, the adversary can wait for the best timing to orchestrate an attack using these keys. He may choose the peak hours of the power usages to send fraudulent commands shut down power generators and create big load mismatch in the power grid. With DKMS, he can only sends out the fraudulent commands right after the intrusion because dynamic keys are updated frequently. The stolen keys are going to expire very soon. Considering the skills and the cost that are necessary to break the security protection of the grid control center, such a short time window to use the stolen keys makes the attack unattractive.

Chapter 6
Secrecy in Communications

Communication security as a research domain is distinctively different than many engineering research disciplines. A research result in communication security may not have constructive performance, i.e. make good things happen. Its functionality is to prevent bad things from happening. In order to present a communication security scheme, the presenter not only needs to show how it works, in which conditions it works, but also to prove why it works.

One goal of this chapter is to explore dynamic secrets from a theoretic point of view. We use a Venn diagram to illustrate the difference between dynamic secrets and the traditional perception of a secret key. The Venn diagram analysis shows that the communication process itself can be a rich source of secrecy, which can be used to persistently reinforce the key strength. Traditional cryptographic key management generally ignore the potential benefits of using communication traffic as a secrecy source. Dynamic secrets make use of this secrecy source and keep refilling secrecy to a dynamic key. As a result, dynamic keys are more resilient to key cracking and key stealing attacks than traditional secret keys.

Another goal of this chapter is to establish connections between dynamic secrets and various other theoretical research domains. The idea of dynamic secrets may be compared with the ideas of quantum key distribution and one-time pad encryption. We also explore the possibility of extends the usage of dynamic secrets to other research problems such as long distance cryptographic key establishment.

The discussion in this chapter involves information theoretic measures such as Shannon entropy and Rényi entropy. We include intuitive interpretations to these measures in this chapter and leave the formal definitions in the appendix.

Section 6.1 challenges the conventional perception of secrets using a Venn diagram analysis. The diagram shows that all the information Alice and Bob share but is unknown to Eve can be used as shared secrets between Alice and Bob. The traditional cryptographic key only accounts for deliberately generated secret information, which is a tiny portion of all available secret information. Dynamic secrets make an efficient use of this secret information and therefore improves communication security.

S. Xiao et al., *Dynamic Secrets in Communication Security*,
DOI: 10.1007/978-1-4614-7831-7_6,
© Springer Science+Business Media New York 2014

Section 6.2 raises the question of how to convert secret information into a secret key. Our answer to this question links the hash function $f_H(\cdot)$ used to generate dynamic secrets with universal random hashing, which is a technique used in the computer database design and in information theoretic security research.

Section 6.3 compares dynamic key updates in a noisy communication environment with quantum key distribution (QKD). The comparison shows that the theoretic foundation of QKD is similar to that of dynamic key updates. We can view dynamic key updates as a low cost, widely applicable implementation of the theoretical idea behind QKD. The trade-off is that QKD delivers an absolutely perfect secret key while dynamic key updates can only produce less-than-perfect key secrecy.

Section 6.4 estimates the difficulty level for an adversary to induce the current value of a dynamic key from a previously obtained key value. Our analysis connects dynamic secrets with Shannon's one time pad encryption, which is the strongest encryption in theory. The result shows that, as time elapses, Eve's key induction would be as difficult as breaking the one time pad encryption.

Section 6.5 relates dynamic secrets to channel coding theory. It is desirable to find a channel coding scheme that is sensitive in detecting transmission errors but inefficient in recovering these errors. We show that previous channel coding research already discovered the existence of such codes. However, these codes are rarely studied because there exist other codes that are better able to improve communication efficiency. Dynamic secrets may revive the study of error detectable but noncorrectable codes and further stimulate the research on the fundamental trade-off between communication efficiency and communication security.

6.1 Adversary's Unknown Information and Secrets

We begin our discussion with a Venn diagram, which represents the general relationship among Alice's, Bob's, and Eve's knowledge. As shown in Fig. 6.1, $I_A(t)$ denotes the set of Alice's knowledge at time t. $I_B(t)$ and $I_E(t)$ are the corresponding knowledge sets for Bob and Eve. The overlapping areas of these knowledge sets represent common knowledge among them. In later discussions, $I_A(t)$, $I_B(t)$, and $I_E(t)$ are also used to denote the binary string forms of these knowledge sets. For example, $I_A(t)$ is a knowledge set. It is also the shortest binary string that can represent the information contained in the knowledge set $I_A(t)$.

In order to quantitatively analyze information and secrets, we must consult with Shannon entropy measure $H(\cdot)$ and Shannon conditional entropy measure $H(\cdot|\cdot)$. $H(X)$ measures the randomness contained in a random variable X, in another word, the uncertainty of X when we treat X as a piece of information. Suppose X is a discrete random variable that takes its value in the value set $\mathscr{X}$. X follows a probability mass function $p_X(\cdot)$. We have

$$H(X) = - \sum_{x \in \mathscr{X}} p_X(x) \log_2 p_X(x). \tag{6.1}$$

Fig. 6.1 A Venn diagram for Alice's, Bob's, and Eve's knowledge sets at time t

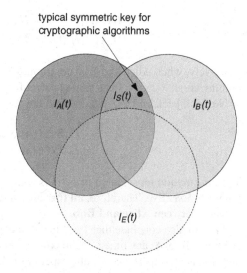

typical symmetric key for cryptographic algorithms

$H(X)$ computed by Eq. (6.1) has a unit of **bit** because the log function is base-2. $H(X)$ measures the level of randomness of X. The more random and uncertain X is, the larger value $H(X)$ is.

Suppose both X and Y are discrete random variables that follow probability mass function $p_X(\cdot)$ and $p_Y(\cdot)$ respectively. They are associated with a joint probability mass function $p_{X,Y}(\cdot, \cdot)$. We have

$$H(Y|X) = - \sum_{x \in \mathcal{X}} \sum_{y \in \mathcal{Y}} p_{X,Y}(x, y) \log_2 p_{Y|X}(y|x) \qquad (6.2)$$

$H(Y|X)$ is the randomness level of Y when X is given as a condition. Naturally, we always have $H(Y|X) \leq H(Y)$ with the equality only occurs when X and Y are independent, i.e. fixing X does not reduce the "freedom" of Y. When we interpret X and Y as two pieces of information. $H(Y|X)$ means when information X is disclosed, how much uncertainty is left for information Y.

In the security context, when X represents a person's knowledge, typically the adversary's knowledge, then $H(Y|X)$ measures the uncertainty of Y to the adversary. Notably, $H(Y|X)$ measures the secrecy level of random variable Y against an adversary whose knowledge is represented by X. If Y represents the cryptographic key, then $H(Y|X)$ is the key secrecy. $H(Y|X) = H(Y)$ means the adversary knows nothing about the key. $H(Y|X) = 0$ means the key is a certain value from the adversary's point of view, i.e. adversary explicitly knows the key.

Using these information entropy measures, we can define a shared secret $x(t)$ for Alice and Bob in the presence of Eve. $x(t)$ is a shared secret if and only if the following two equations holds.

$$H(x(t)|I_A(t)) = H(x(t)|I_B(t)) = 0 \tag{6.3}$$
$$H(x(t)|I_E(t)) = H(x(t)) \tag{6.4}$$

Intuitively, when Alice and Bob commonly know the value of $x(t)$ and Eve knows nothing about $x(t)$, $x(t)$ is a shared secret for Alice and Bob against Eve.

We look back into Fig. 6.1 and find the set of shared secrets, $I_S(t)$.

$$\begin{aligned} I_S(t) &= \{x(t)|x(t)\textbf{is a shared secret}\} \\ &= I_A(t) \cap I_B(t) - I_E(t) \end{aligned} \tag{6.5}$$

The information represented in $I_S(t)$ is commonly known to Alice and Bob but not known to Eve, Eve. Therefore, **all the information contained in $I_S(t)$ is the shared secrecy between Alice and Bob**.

We can also conclude that $I_S(t)$ is the universal set of all shared secrets between Alice and Bob. At any time t, the maximum amount of shared secrecy between Alice and Bob is $|I_S(t)|$. The cardinality operator $|\cdot|$, when applied on an information set, calculates the smallest number of bits needs to represent the information contained in the information set.

The set of shared secrecy, $I_S(t)$, contains two types of secret information. The first type of secrets are those pre-shared secret information, such as the traditional secret key configured by a security administrator before Alice and Bob communicate. This type of secret information also includes the secrets shared through a pre-secured channel such as the key established with the help of a quantum key distribution system. This type of secret information requires the help of third party, either a trusted administrator or a secured channel.

Another type of secret information is **the information exchanged between Alice and Bob in a public channel but missed by Eve**. The second type of secret information is spontaneously generated during the course of communication. The generation of these secrets do not need any third party help.

Let Alice and Bob begin to communicate at time $t = 0$. Assuming that there is no third party to help generate and distribute the shared secrets between Alice and Bob for $t > 0$, then $I_S(t)$ is divided into two disjoint subsets.

$$I_S(t) = I_S(0) \cup I_S^\delta(t) \tag{6.6}$$

$I_S(0)$ is the subset of the secret information that Alice and Bob shared prior to any communication between them. $I_S^\delta(t)$ represents the subset of information exchanged between Alice and Bob through communications but missed by Eve.

Even if we allow Eve to eavesdrop the Alice-Bob communication with minimal risk of being caught, it is often difficult for Eve to maintain an everlasting, perfect eavesdropping. Practical constraints such as budget limit and durability of eavesdropping devices prevent an ideal, everlasting signal eavesdropping from being impractical. When Eve is not eavesdropping, all the information exchanged between Alice and Bob naturally becomes their shared secrets.

When Eve eavesdrops, the randomness in the communication environment will result in her information loss, which becomes Alice's and Bob's shared secret as well. In practice, $I_S^\delta(t)$ is often the dominant part of $I_S(t)$. Moreover, $I_S^\delta(t)$ grows over time. As the communication continues, "fresh" secret information is continually added to $I_S^\delta(t)$ and becomes available to the users. It is a great waste if we only use the secret in $I_S(0)$ but ignore the rich secrecy contained in $I_S^\delta(t)$.

6.1.1 Traditional Symmetric Key

The traditional symmetric key only makes use of deliberately generated secret information contained in the information set $I_S(0)$. The amount of secrecy is inherently limited. The amount of key secrecy for traditional symmetric keys, which is typically 128 bits or 256 bits, are much less than the amount of shared secrecy contained in $I_S^\delta(t)$ for large t in many practical secure communication scenarios.

In practice, Alice and Bob often do not have access to third party help for time $t > 0$. They can change the value of their symmetric key but they cannot improve the key secrecy. Alice and Bob can update their symmetric key by agreeing on a new key value through traditional key update protocols (e.g., [89]) . However, traditional key update protocols require a previously existing key to secure the key exchange process. The new key's secrecy relies on the secrecy of the previous key. From an information security point of view, the new key is merely a derived key from the previous key. It is the secrecy of previous key that secure the communication between Alice and Bob although they may use the new key for encryption and message integrity check.

6.1.2 Asymmetric Key

Asymmetric key cryptography assigns a pair of keys for each user, namely k_{pub} and k_{prv}. The public key k_{pub} is released as public knowledge and the private key k_{prv} is securely kept by the user. When other users want to send a message securely to the owner of the asymmetric key, they use k_{pub} to encrypt the message, and then the key owner decrypts the message using k_{prv}.

Asymmetric key cryptography allows users to securely exchange messages without any pre-shared secret. However, k_{prv} can be calculated from k_{pub} if Eve has sufficient computing power. Therefore the messages encrypted by asymmetric key cryptography are not secret from an information theoretic perspective. Alice and Bob cannot generate shared secrets using asymmetric key cryptography.

Even if we assume Eve does not have enough computing resources to calculate k_{prv} from k_{pub}, asymmetric key cryptography still has two pitfalls. Firstly, the computational costs of asymmetric key encryption and decryption is high. In practice, users can only afford to use asymmetric key cryptography to exchange a symmetric key and then use the symmetric key to establish secure communication. The secrecy of this symmetric key depends on the secrecy of k_{prv}. As a consequence, the communication security completely relies on the safety of k_{prv}. Secondly, because k_{pub} is publicly known, its authenticity must be endorsed by a trusted third party. Otherwise Eve can release fake public keys to impersonate as Alice or Bob. The setup and maintenance costs for the trusted third party infrastructure is expensive for many secure communication applications.

Dynamic secrets can be used to complement asymmetric key cryptography. Alice and Bob establish an initial symmetric key through an asymmetric key secured key agreement process. Then dynamic secrets begin to update the symmetric key using the secrecy in communications. We have discussed about this possibility in the last section of Chap. 3.

6.1.3 Dynamic Secrets and Dynamic Key

Dynamic secrets, from an information theoretic point of view, gain their secrecy from $I_S^\delta(t)$. Suppose a sequence of dynamic secrets $s(\tau_i)$ is generated at times τ_i, $i = 1, 2, \ldots$. The information communicated between Alice and Bob in the time period $[\tau_{i-1}, \tau_i)$ is denoted as $m(\tau_i)$. $s(\tau_i)$ is generated by hashing $m(\tau_i)$,

$$s(\tau_i) = f_H(m(\tau_i)). \tag{6.7}$$

We divide $I_S^\delta(t)$ into a sequence of disjoint subsets $I_{S,sub}^\delta(\tau_i)$,

$$I_{S,sub}^\delta(\tau_i) = I_S^\delta(\tau_i) - I_S^\delta(\tau_{i-1}) \tag{6.8}$$

for $i = 1, 2, \ldots$ and $\tau_i < t$. Here we use $-$ as set difference operator. For each time period $[\tau_{i-1}, \tau_i)$, the shared secret information, i.e. the shared secrecy, is a subset of the information exchanged in that time period,

$$I_{S,sub}^\delta(\tau_i) \subseteq \{m(\tau_i)\}. \tag{6.9}$$

Since $s(\tau_i)$ is the hash value, i.e. a function mapping, of $m(\tau_i)$, $s(\tau_i)$ carries secrecy from $I_{S,sub}^\delta(\tau_i)$. When Eve completely eavesdrops all the bits in $m(\tau_i)$, we have $I_{S,sub}^\delta(\tau_i) = \phi$ and $s(\tau_i)$ can be calculated by Eve. However, if Eve encounters decoding errors or packet losses when eavesdropping $m(\tau_i)$, the chance that Eve cannot calculate $s(\tau_i)$ is overwhelmingly high.

The dynamic key, $k(t)$, is a symmetric key between Alice and Bob. Whenever a dynamic secret is generated, Alice and Bob update the key by XORing it with the dynamic secret.

$$k(\tau_i) = k(\tau_{i-1}) \oplus s(\tau_i). \tag{6.10}$$

When Eve has uncertainty about $s(\tau_i)$, she would be more uncertain about $k(\tau_i)$ than $k(\tau_{i-1})$. From users' point of view, the secrecy carried by $s(\tau_i)$ joins the key secrecy. From time 0 to time t, the sequence of dynamic secrets generated in this time period makes a full use of the secret information set $I_S^\delta(t)$. As communication goes on, dynamic secrets persistently feed shared secrecy from Alice and Bob's communication to the key.

6.1.4 Man-in-the-Middle Adversary with Key

At the end of Chap. 3, we have mentioned the Man-In-The-Middle (MITM) adversary, i.e. an adversary who sits between Alice and Bob and can intercept and relay *all* of the communications. An MITM adversary is able to deduce all the dynamic secrets and track all the dynamic key updates. Dynamic secrets do not provide stolen key recovery against an MITM adversary. When an MITM adversary obtains the key, communication security fails and does not recover even if dynamic secrets have been used.

Figure 6.1 intuitively explains the invincibility of an MITM adversary with knowledge about the cryptographic keys. $I_S(t)$ contains all the information that can be used as shared secrets for Alice and Bob. As analyzed above, $I_S(t)$ can be divided into two subsets, $I_S(0)$, which contains all the pre-shared secrets and $I_S^\delta(t)$, which contains all the secret information generated by the Alice-Bob communication.

When the MITM adversary knows all the cryptographic keys, e.g. the symmetric key and the private key k_{prv} in the public key pair, $I_S(0)$ has no secret to Eve, i.e.

$$I_S(0) = \phi. \tag{6.11}$$

The MITM adversary intercepts and relays all the messages exchanged between Alice and Bob. Therefore we have

$$I_S^\delta(t) = \phi. \tag{6.12}$$

It is impossible for Alice and Bob to gain any secrets through their communications, either using dynamic secrets or any other methods.

As a result of Eqs.(6.11) and (6.12), we have $I_S(t) = \phi$ for all t. Alice and Bob do not have any shared secrets to support the cryptographic protections at any time. Please note that this pessimistic conclusion works for both symmetric key and asymmetric key case when all the user kept secrets are known to the adversary.

6.2 Universal Hashing and Secrecy Extraction

We have discussed about the importance of using all the secret information in $I_S(t)$ instead of only using $I_S(0)$. It is time to look for an efficient technology to utilize $I_S(t)$. Although Alice and Bob know that all the information contained in $I_S(t)$ can be used as their shared secrets, it is not straightforward for them to extract and use the secret information because Eve's knowledge set, $I_E(t)$, is unknown. Alice and Bob cannot distinguish shared secret information in $I_S(t)$ from the information commonly known by Alice and Bob and known to Eve, i.e. $I_A(t) \cap I_B(t) - I_S(t)$.

In order to efficiently extract secrecy from $I_A(t) \cap I_B(t)$, Alice and Bob can use a technique known as *universal-2 hashing*. Universal-2 hashing was first introduced in [25] as a technique to minimize collisions in hash computations. For the inputs with an arbitrary probability distribution, if the hash function is uniformly randomly chosen from a class of universal-2 hash functions, the distribution of the hash outputs will be as close to the uniform distribution as possible. The formal definition of universal-2 hash functions can be found in Appendix A.

Universal-2 hashing has been used to extract perfect secrets from non-perfect secret resources in information theoretic secrecy research [22, 83]. We can apply a similar approach here to maximize the efficiency of secrecy extraction from $I_A(t) \cap I_B(t)$.

Let $I_{AB}(t) = I_A(t) \cap I_B(t)$ denote Alice's and Bob's common information at time t. If Eve only has incomplete information about $I_{AB}(t)$, then from Eve's view, the possible values of $I_{AB}(t)$ is random with probability distribution $p(I_{AB}(t)|I_E(t))$, which could be an arbitrary distribution.

Alice and Bob apply universal-2 hashing on $I_{AB}(t)$ and generate the secret $S(t) = f_{UH}(I_{AB}(t))$. According to the properties of universal hashing, regardless of the actual input conditional distribution $p(I_{AB}(t)|I_E(t))$, the output conditional distribution $p(S(t)|I_E(t))$ will be as close to a uniform distribution as possible. We apply the Shannon entropy measure to the universal hash output. The closer the distribution $p(S(t)|I_E(t))$ is to the uniform distribution, the higher the value $H(S(t)|I_E(t))$ is, i.e. $S(t)$ contains more secrecy. Because universal-2 hashing generates $S(t)$ from $I_{AB}(t)$ and makes $S(t)$ as close to a uniform distribution as possible, we can conclude that universal-2 hashing maximally retains secrecy from Alice's and Bob's common knowledge set, $I_{AB}(t)$.

6.2.1 Secrecy Extraction Efficiency

In order to analyze the secrecy extraction efficiency, we need to use both the Shannon entropy measure and the Rényi entropy measure of order 2 [104], denoted as $H(\cdot)$ and $H_2(\cdot)$ respectively. Let X denote a discrete random variable. It is always true that $H(X) \geq H_2(X)$ and $H_2(X) > 0$ when $H(X) > 0$. A more detail discussion about these two information entropy measures can be found in Appendix B.

The secrecy of a random variable is defined by its Shannon conditional entropy given Eve's knowledge. Let X be a random variable. I_E is a random variable that represents Eve's knowledge. X's secrecy is measured by $H(X|I_E)$, e.g. the amount of X's uncertainty from Eve's point of view [113]. Therefore, the amount of secrecy contained in $I_S(t)$ is measured as

$$|I_S(t)| = H(I_{AB}(t)|I_E(t)). \tag{6.13}$$

However, Alice and Bob cannot extract $|I_S(t)|$ bits of secret information because they have no information about Eve's knowledge set, $I_E(t)$. The amount of secret information that can be extracted is measured by $H_2(I_{AB}(t)|I_E(t))$. The difference between $H(I_{AB}(t)|I_E(t))$ and $H_2(I_{AB}(t)|I_E(t))$ is the price paid for the blind secrecy extraction process.

Theorem 6.1 *(Theorem 3 in [21]) Let $\mathscr{F} : GF(2^{l_{AB}}) \rightarrow GF(2^{l_S})$ be a class of universal-2 hash functions where l_{AB} is the number of bits in $I_{AB}(t)$ and l_S is the binary length of $S(t)$. Let $f_{UH}(\cdot)$ be a uniformly randomly chosen function from $\mathscr{F}$. $S(t)$ is defined by*

$$S(t) = f_{UH}(I_{AB}(t)). \tag{6.14}$$

Suppose $H_2(I_{AB}(t)|I_E(t)) \geq \varepsilon > 0$ where ε is an arbitrary positive number, and the random variable F represents the uniformly random choice of hash function from $\mathscr{F}$, the following inequality holds,

$$H(S(t)|F, I_E(t)) \geq l_S - \log_2(1 + 2^{l_S - \varepsilon}). \tag{6.15}$$

It's noteworthy that the result in (6.15) is averaged over all uniformly random choices of hash functions in $\mathscr{F}$. There is a non-zero probability that when $H_2(I_{AB}(t)| I_E(t)) \geq \varepsilon$, for some specific values of F, the secrecy in $S(t)$, i.e. $H(S(t)|I_E(t))$, breaks the lower bound in (6.15). However, such combinations of $I_{AB}(t)$, F and $I_E(t)$ are extremely rare and can be ignored in practice [21].

Theorem 6.1 is used in the quantum key distribution [19] for the purpose of privacy amplification, i.e. convert a long, partially secret binary string into a short, perfectly secret string. This theorem is also used in many other information theoretic security schemes [53] to perform similar tasks. Here we expand this theorem into three corollaries for our purpose of efficiently extracting secrecy from $I_{AB}(t)$ when $I_E(t)$ is unknown.

Corollary 6.1 *If $l_S < \varepsilon$, Eve knows less than one bit of information of $S(t)$,*

$$H(S(t)|F, I_E(t)) \geq l_S - \log_2(1 + 2^{-x}) > l_S - 1. \tag{6.16}$$

Corollary 6.2 *If $l_S \geq \varepsilon$, $S(t)$ contains at least $\varepsilon - 1$ bits of secrecy,*

$$H(S(t)|F, I_E(t)) \geq l_S - (1 + \log_2 2^{l_S - \varepsilon}) \geq \varepsilon - 1. \tag{6.17}$$

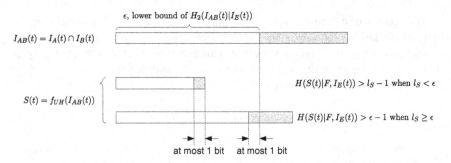

Fig. 6.2 An illustration for corollaries 6.1 and 6.2

Corollary 6.3 *If the value of* $H_2(I_{AB}(t)|I_E(t))$ *follows a probability distribution such that*

$$p(H_2(I_{AB}(t)|I_E(t))) \geq \varepsilon > 0) \geq 1 - \delta, \tag{6.18}$$

then the secrecy contained in $S(t)$ *is lower bounded as*

$$H(S(t)|F, I_E(t)) \geq (1 - \delta)(l_S - \log_2(1 + 2^{l_S - \varepsilon})). \tag{6.19}$$

As shown in Fig. 6.2, Corollaries 6.1 and 6.2 together demonstrate the secrecy extraction efficiency when $H_2(I_{AB}(t)|I_E(t))$ is lower bounded by a positive number ε. If $S(t)$ is not long enough to contain ε bits of secrecy, $S(t)$ will be an almost perfect secret, with less than one bit of information known to Eve. If $S(t)$ is longer than ε bits, then it will contain at least $\varepsilon - 1$ bits of secrecy.

In the wireless communication scenarios, Corollary 6.3 is particularly useful. Alice and Bob can estimate Eve's reception errors using the statistical channel models such as the additive white Gaussian noise channel model or the Rayleigh fading channel model. Then the value of $H_2(I_{AB}(t)|I_E(t))$ can be described by a probability distribution. The users can estimate the secrecy of $S(t)$ using Corollary 6.3.

Moreover, Alice and Bob can intentionally send wireless jamming signals to disrupt the eavesdropping channel and ensure either a constant lower bound or a probabilistic lower bound of Eve's information loss [45]. Then they can apply the above corollaries to estimate the quality of secrets generated from their communications.

6.2.2 Deterministic Hashing and Mixed, Imperfect Secret

Although universal-2 hashing provides a theoretically efficient method for secrecy extraction, it is hardly practical. Universal-2 hash functions are computationally expensive and require dedicated hardware acceleration to be feasible for real world computer systems [7, 67]. We may use a deterministic, cryptographic hash function $f_H(\cdot)$ as a practical alternative to the universal-2 hash function, $f_{UH}(\cdot)$.

Suppose $S(t) = f_H(I_{AB}(t))$ and $f_H(\cdot)$ is a deterministic hash function with cryptographic strength. If Eve only has incomplete information about $I_{AB}(t)$, then $S(t)$ will be a mixture of secrets and the information known to Eve. It is possible that Eve only has incomplete information about $I_{AB}(t)$, but the deterministic hash function help eliminate Eve's uncertainty by mapping all the possible values of $I_{AB}(t)$ conditioned on Eve's knowledge into the same hash value. The odds of such case in practice is negligible. Moreover, it is the environmental randomness that determines Eve's knowledge about $I_{AB}(t)$. Eve cannot control $I_E(t)$ to adapt the deterministic hash function to gain advantages in predicting $S(t)$.

If the strictest security standard is imposed, $S(t)$ generated by deterministic hashing cannot be used as a cryptographic key. In theory, the values of a cryptographic key must be uniformly distributed in the key space. Otherwise adversary may exploit the probabilistic characteristics of the key to weaken the protection of encryption or keyed messaged integrity check algorithms. Practical secure communication users tend to neglect such a theoretic threat. In most practical cases, a 512 bits long key with 256 bits of secret information is considered to be more secure than a 56 bits perfectly secure key. Furthermore, $I_S^\delta(t)$ could contain a vast number of secret bits and the amount of secret information keeps growing as communication goes on. We argue that it is practically feasible to use deterministic hash for extraction in dynamic secrets based communication security schemes.

When a deterministic hash function is used, the exact ratio of secrecy information contained in $S(t)$ is hard to analyze because of the complex structure of $f_H(\cdot)$. We cannot provide analytic information bounds such as found in Corollaries 6.1, 6.2, and 6.3. However, when $f_H(\cdot)$ is a strong cryptographic hash function, such as SHA-256, it is practically infeasible for Eve to use her knowledge of $I_{AB}(t)$ to infer the exact bits or probabilistic distribution of $S(t)$.

6.2.3 Secrecy Extraction in Small Time Segments

We have to consider another practical constraints in secrecy extraction, the storage limit. The communication relationship between Alice and Bob could last for years. It is impractical for Alice and Bob to store all the messages they exchanged from time 0 to a large time t and hash all these messages to get a secret $S(t)$ as their shared secret key $k(t)$.

Dynamic secrets generation uses $f_H(\cdot)$ iteratively to reduce the amount of storage. The large time period $[0, t)$ is divided into a sequence of small time periods, $[0, \tau_1), [\tau_1, \tau_2), \ldots$ At time τ_i, a dynamic secret $s(\tau_i)$ is generated by hashing the information exchanged between Alice and Bob during the time period $[\tau_{i-1}, \tau_i)$. Then the stored information is discarded. Alice and Bob only need to provision a buffer large enough to hold the information exchanged in each small time period.

6.3 Quantum Key Distribution and Dynamic Key

It is quite natural to transit the discussion from universal-2 hashing to quantum key distribution (QKD) because universal-2 hashing is widely used in the privacy amplification step of QKD. Not only dynamic secrets share an important technique with QKD, but also their ideas are fundamentally similar.

6.3.1 QKD in a Dynamic Secrets View

The purpose of the quantum key distribution is to establish a perfectly secret key between Alice and Bob. Taking the BB84 protocol as an example [20], Alice first generates a random string of bits and for each bit she randomly chooses one of two sets of orthogonal bases to encode the bit, as shown in Fig. 6.3.

Alice transmits a sequence of polarized photons to Bob. Each photon represents a bit and the polarization direction of the photon is the same as the bit's encoded direction. Bob also randomly chooses one of the two sets of bases, diagonal or vertical-horizontal, to measure each incoming photon. He would expect to have a 50 % chance to correctly decode these photons. After Bob receives and measures all photons. He publicly discusses the basis choices for each measurement with Alice. They then locate Bob's correct measurements and keep these measurement results as their shared secret key. An example BB84 quantum key distribution process is shown in Fig. 6.4.

The quantum channel has a unique property that allows Alice and Bob to detect the eavesdropping adversary. In theory, Eve cannot eavesdrop any information. 50 % of the bits sent by Alice will become the shared secrets between Alice and Bob. The Venn diagram in Fig. 6.1 can be used to interpret the quantum key distribution process. At time 0, Alice and Bob does not share any information. At time t, Alice and Bob shares 50 % of the bits sent by Alice. These shared bits form the knowledge set $I_A(t) \cap I_B(t)$. Because Eve is not allowed to eavesdrop, his knowledge set $I_E(t)$ has no intersection with $I_A(t) \cap I_B(t)$. Therefore the shared secrecy set

$$I_S(t) = I_A(t) \cap I_B(t) - I_E(t) = I_A(t) \cap I_B(t). \tag{6.20}$$

basis \ bit	0	1
$\times$ (d)	↗	↘
$+$ (v)	↑	→

Fig. 6.3 The BB84 encoding scheme for a quantum bit (qubit). There two sets of bases: diagonal (d) and vertical-horizontal (v). The actual polarization direction of the transmitted photon is denoted by the arrows

Alice's random bits	0	1	1	0	1	0
Alice's basis choices	×	+	×	×	+	+
Photon polarizations	↗	→	↘	↗	→	↗
Bob's basis choices	+	+	×	+	×	+
Public discussion		√	√			√
Secret key bits		1	1			0

Fig. 6.4 An example process of BB84 protocol. Alice and Bob agrees on 3 bits as their shared secret key

All the bits contained in $I_S(t)$ can be used as shared secrets between Alice and Bob. This explains the outcome of BB84 protocol.

We may further allow Eve to break the physics barrier and eavesdrop the quantum channel without disturbing the photons such as using the weak measurement technique [107]. As long as Eve can only measure each photon once and his basis choices are independent of Bob's basis choices, Eve can expect to know 50 % of the bits that Bob correctly decodes, which accounts for 25 % of bits sent by Alice. In this case, the shared secrecy set $I_S(t)$ contains 25 % of the bits sent by Alice. Bob can apply the universal-2 hash function over the shared information set $I_A(t) \cap I_B(t)$ and produce a close-to-perfect secret that has a binary length of $|I_A(t) \cap I_B(t)|/2$.

Just like dynamic secrets, quantum key distribution makes use of $I_S(t)$, the shared secrecy established by communicating in a "noisy" channel. The quantum channel can be viewed as a highly random channel where Eve suffers 100 % information loss without the weak measurement technique or 50 % information loss if weak measurements are allowed. Eve's information loss is the user's shared secrecy.

6.3.2 Dynamic Secrets as Low Cost QKD

When Alice and Bob communicate through the quantum channel, they can generate perfectly secret bits at a high rate because the channel forces Eve to suffer a high bit loss probability. However, the cost to set up and maintain a quantum channel is exorbitant. Moreover, the engineering imperfections in practice substantially degrade the performance of the quantum key distribution. We can use dynamic secrets as a low cost, low secrecy rate alternative to the quantum key distribution in a non-quantum communication channel. Although Eve may have lower bit loss probability when eavesdropping in a non-quantum communication channel, the low communication cost enables Alice and Bob to exchange many more bits than they can transmit in the quantum channel. The large volume of communication traffic will compensate for the reduction in Eve's information loss rate and provide a sufficient amount of shared secrecy to secure the communication between the users.

We argue that in most typical communication scenarios, channel randomness is sufficient for the dynamic secrets to provide a strong protection against key theft and an eavesdropping adversary. In these cases, we can view dynamic secrets generation and dynamic key updates as a low cost quantum key distribution. The channel noise acts as the weak quantum indeterminacy that produces unavoidable information loss to Eve.

Wireless communication is known for error-prone in signal transmissions. [112] reports that 2G and 3G telecommunications have a typical packet loss ratio between 0.05 and 0.1. WiFi channels are heavily affected by the physical environment and its utilization. In a laboratory environment, [27] reports 0.01–0.05 packet loss for light traffic load. In close range wireless communications, 802.15.4 low rate personal wireless network is of wide interests from both the industry and the academia. A 0.1 level of packet loss ratio in an industrial environment by [33].

The above mentioned packet loss ratio is measured with typical receivers that the users would have. Eve could have better receivers to reduce his packet losses. However, packet losses are caused by multiple random factors such as thermodynamic noise in the receiver, noise in the signal propagation environment, multi-path interference and interferences from other signal sources. It is infeasible to build a wireless receiver that can suppress all the random factors in most realistic communication environments.

Wired communication is less random than the wireless communication. However, random factors exist in every layer of the multiple layer design. The combined effect of these random factors would be non-negligible and provide helpful influence to communication security when dynamic secrets are used. As reported in [54], the physical layer noise in a telephone line may corrupt 10 % of packets flow trough it. In the network layer, the Internet router may discard packets according to the traffic utilization [40]. Such packet losses are unfavorable for Eve.

Internet routing dynamics is also a source of communication randomness. [94] reports that 9 % of Internet paths change in the order or 10 s of minutes and 50 % of routes persist for less than seven days. An adversary would find it difficult to tap an end-to-end communication giving such routing dynamics. In the future, the deployment of multi-path data transmission [59, 60] may further increase communication randomness and result in Eve's information loss.

In general, wired communications are more favorable for Eve to eavesdrop than wireless communications. A worst case scenario is that Eve completely controls an access router of either Alice or Bob. Eve would be able to know every packet exchanged between Alice and Bob and calculate all key updates. However, this scenario is barely possible in theory. The communication between Alice and Bob may last months and years in practice. It is practically impossible for Eve to control the router before Alice and Bob ever communicate with each other and then keep a complete control of the router for months and years.

6.3.3 Long Distance Key Establishment

An important usage of QKD is to establish cryptographic keys between military deployments such as between two satellites or two remote sites that locate hundreds of kilometers away from each other. It is technically challenging to transport quantum entangled photons across hundreds or even thousands of kilometers. As far as this monograph is written, the record of long distance quantum key distribution is held by Chinese scientist who have successfully demonstrated free space transportation of quantum entangled photons in 120 km [132]. They are planning a free space experiment between a near orbit satellite and a ground base which could extends the distance record to several thousand kilometers. Needless to mention, the experiment devices cost millions of dollars.

We may use dynamic secrets to establish cryptographic keys in long distance at lower cost. The transmission technology of radio signals over thousands of kilometers have been invented for more than a hundred years since there was telegraph. Further, because the advancement of space exploration and the development of astronomy, we are able to communicate between Mars rovers and Earth control center [17] in a distance of at least 55,000,000 kilometers. There is no technical challenge to send and receive radio signals between two distantly located sites.

When radio communications are available, we can bootstrap a symmetric key between two sites using dynamic secrets as discussed in Chaps. 3 and 4. There are two differences between long distance QKD and dynamic secrets based long distance key establishment. QKD can output key bits instantly while dynamic secrets require a period of communication to bootstrap the key to a satisfactory level of security strength. The key bits come from QKD are perfectly secure while the key bootstrapped by dynamic secrets is secure with high probability.

6.4 Secrecy Sharing, Perfect Secret, and Dynamic Secrets

Although dynamic secrets are mainly a practical vehicle to exploit traditionally unused secrecy source for communication security, the idea of dynamic secrets has close relationship with various information theoretic security researches. One of the closest relatives is the research on secrecy sharing problem.

6.4.1 Secrecy Sharing Problem

There are many research works focus on how to generate a perfectly secret, shared key for Alice and Bob, when Alice and Bob could communicate with each other but do not share any secrets beforehand. The research problem targeted by these research works is named as secrecy sharing problem.

In wireless communications, it is often a realistic assumption that Eve's eavesdropping channel is independent to the communication channel between Alice and Bob. Therefore any physical property of the communication channel is a perfect secret to Eve. Notably, Alice and Bob can measure their channel reciprocity and use the measurement result as their shared secret. Reference [8, 10, 47, 82, 127] are theoretic and experimental research works using this channel reciprocity idea to generate shared secret key.

Reference [14] suggests that one could to exploit independent channel fades for opportunistic secrecy sharing. Suppose Eve's eavesdropping channel is in deep fading and at the same time Alice and Bob have a communication channel with higher instantaneous capacity. The information exchanged between Alice and Bob at this moment could be converted into a perfect shared secret for them.

By assuming that Alice and Bob have perfect knowledge about their communication channel, [24, 25] propose to use capacity achieving codes for secrecy sharing. Reference [129] proposes Detectable Non-Correctable (DNC) codes for the same objective. The channel knowledge assumption is relaxed at the cost of secrecy sharing efficiency. References [121, 128] combine channel coding with reliable communication mechanisms to share secrets for Alice and Bob. Reference [77, 78] propose to use antenna diversity to disseminate secrets. These approaches yield unconditional security for the shared secrets. However, they either demand perfect channel measurements or require special hardware modifications, and therefore, are not generally applicable to practical wireless communications.

A more general framework for secrecy sharing problem has been investigated by Maurer [84, 85, 86]. In Maurer's work, the users' knowledge and Eve's knowledge are modeled as correlated random variables. The conditions under which users can share information theoretic secrets are obtained. Maurer's approach involves the use of random hashing techniques such as universal hash functions [26] and random extractors [28], that enable analytic calculations for the achievable rate of shared key generation but make the framework highly theoretical.

References [70, 71, 92] attempt to convey the information theoretic security into practical wireless applications. They focus on physical layer secrecy capacity with ARQ retransmissions and improve security in the wired equivalent privacy (WEP) protocol.

The goal of dynamic secrets is different. There is no guarantee that any dynamic secret will be a perfect secret. Instead, dynamic secrets are made up of many imperfect secrets. We propose to XOR many imperfect secrets during the communication process to provide a persistent and ever-strengthening security.

6.4.2 Non-decreasing Key Secrecy

Recall the notations used in Chap. 3. $m(\tau_i)$ for $i = 1, 2, \ldots$ denote messages exchanged between Alice to Bob in time periods $[\tau_{i-1}, \tau_i)$. Suppose Alice and Bob exchange sufficiently independent messages such that $m(\tau_i)$ for $i = 1, 2, \ldots$ are mutually independent conditioned on Eve's knowledge.

$$
\begin{aligned}
p(m(\tau_1), &\ldots, m(\tau_{i-1}), m(\tau_i)|I_E(\tau_i)) \\
&= p(m(\tau_1), \ldots, m(\tau_{i-1})|I_E(\tau_i))p(m(\tau_i)|I_E(\tau_i))
\end{aligned}
\tag{6.21}
$$

When Eq. (6.21) holds, Eve cannot recover her prior information loss by eavesdropping later messages. Dynamic secrets are hash values of $m(\tau_i)$ for $i = 1, 2, \ldots$ and are also conditionally independent.

$$
\begin{aligned}
p(s(\tau_1), &\ldots, s(\tau_{i-1}), s(\tau_i)|I_E(\tau_i)) \\
&= p(s(\tau_1), \ldots, s(\tau_{i-1})|I_E(\tau_i))p(s(\tau_i)|I_E(\tau_i))
\end{aligned}
\tag{6.22}
$$

The key secrecy can be expressed as a Shannon entropy measure.

$$
S_k(\tau_i) = H(k(\tau_i)|I_E(\tau_i))
\tag{6.23}
$$

The relationship between two successive keys is

$$
k(\tau_i) = k(\tau_{i-1}) \oplus s(\tau_i).
\tag{6.24}
$$

Furthermore, the dynamic key incorporates secrecy that is non-decreasing in time,

$$
\begin{aligned}
S_k(\tau_i) &= H(k(\tau_{i-1}) \oplus s(\tau_i)|I_E(\tau_i)) \\
&\geq H(k(\tau_{i-1}) \oplus s(\tau_i)|I_E(\tau_i), s(\tau_i)) \\
&= H(k(\tau_{i-1})|I_E(\tau_i), s(\tau_i)) \\
&= H(k(\tau_{i-1}|I_E(\tau_{i-1}) \\
&= S_k(\tau_{i-1})
\end{aligned}
\tag{6.25}
$$

In practice, the conditional independence assumption of Eq. (6.21) can be further relaxed. The joint probability distribution of messages is not available in practice. Eve can only use semantic relationship between messages to help infer her lost messages. It is infeasible for Eve to use later eavesdropped messages to deduce the exact binary form of prior messages and compensate for the bit level information loss in his earlier eavesdropping.

Moreover, Eve can only eavesdrop the encrypted form of messages. Once Eve suffers a decoding error while eavesdropping and fails to track the dynamic key updates, she will not be able to decrypt further eavesdropped messages. From a practical point of view, Eve's information loss is irreversible. The accumulation of Eve's information loss will always strengthen the key secrecy. As time goes to infinity, the key secrecy approaches perfect secrecy.

6.4.3 Perfect Secret

Perfect secret is a mathematical concept proposed by Shannon [113]. Using Shannon entropy measure, a random variable X is a perfect secret against adversary's knowledge I_E if and only if

$$H(X|I_E) = H(X). \tag{6.26}$$

This equation means that the adversary cannot reduce any bit of uncertainty of X by using his knowledge. In communication security context, a perfect secret encryption is defined by the conditional entropy between plain text M and its cipher text C. When

$$H(M) = H(M|C), \tag{6.27}$$

the encryption is considered as perfectly secure or information theoretically secure because adversary cannot deduce any information of M by looking at C.

Shannon further proved that, the only way to achieve perfect encryption is using One Time Pad (OTP) encryption or its equivalence. In another word, one time pad encryption provides the theoretically strongest protection for the plain text message M.

6.4.4 Asymptotic One Time Pad

Let x be the plain text to be encrypted. p is a one time pad, i.e. a binary string that is unknown to Eve. c denotes the encrypted outcome, the cipher text. One time pad encryption can be expressed as

$$x \oplus p = c. \tag{6.28}$$

The cipher text c is then transmitted over an insecure channel. Eve may learn c. However, as long as p remains unknown to Eve, it can be proved that Eve learns nothing about x. The one time pad provides the strongest possible encryption protection to the plain text x against Eve.

When dynamic secrets are used, the relationship between $k(t)$, $k(0)$, and the dynamic secret generated during time interval $(0, t]$ can be expressed as

$$k(t) = k(0) \oplus \bigoplus_{0 < \tau_i \leq t} s(\tau_i) = k(0) \oplus s_{(0,t]}, \tag{6.29}$$

where

$$s_{(0,t]} = \bigoplus_{0 < \tau_i \leq t} s(\tau_i). \tag{6.30}$$

Equation (6.29) can be reinterpreted as

$$k(t) \oplus s_{(0,t]} = k(0) \tag{6.31}$$

The structure of Eq. (6.31) is identical to that of Eq. (6.28). $k(0)$ is the result of "encrypting" $k(t)$ with $s_{(0,t]}$. Eve may obtain $k(0)$ through key theft, just like She may learn the cipher texts c by eavesdropping the insecure channel that c goes through. However, Eve will not obtain $k(t)$ because she does not know $s_{(0,t]}$. If $s_{(0,t]}$ is a perfect secret from Eve's point of view, then calculating $k(t)$ from $k(0)$ will be as difficult as breaking one time pad encryption for Eve.

As indicated by Eq. (6.25), the key secrecy is never decreasing. This process, when interpreted from the one time pad perspective, shows that the secrecy of $s_{(0,t]}$ never decreases over time. $s_{(0,t]}$ may not be very secret when t is small. As t grows, the XORing of many not-so-perfect dynamic secrets will push the secrecy of $s_{(0,t]}$ up. As t goes to infinity, $s_{(0,t]}$ approaches an one time pad. Suppose $k(0)$ is known to Eve, she will find that the chances of calculating $k(t)$ becomes smaller and smaller as time goes by.

6.5 Error Detectable but Non-correctable Codes

Dynamic secrets has a connection with channel coding. When messages are sent through a noisy communication channel, channel coding techniques are used to help the receiver detect and correct possible bit errors. The error detection capability of channel codes is necessary for Alice and Bob to identify and retransmit the erroneous messages. The error correction capability may allow Alice and Bob to reduce the message retransmissions and improve the communication efficiency. However, Eve can also exploit the error correction capability and reduce her information loss in eavesdropping.

Suppose we want to design a code that allows Alice and Bob to maximize their information privilege over Eve. It is best to design a channel code that is able to detect errors but unable to recover errors, namely error detectable but non-correctable (DNC) codes. Such channel codes are usually of interest to the communications community because they do not provide the best communication performance.

In this section, we introduce *equiprobable parity check codes* as an example to prove the existence of the DNC codes. [129] first proposes the security usage of the equiprobable parity check codes over a binary symmetric channel (BSC).

Suppose the message to be coded is a row vector of nR bits, denoted as $\mathbf{t_m}$. The generation matrix of equiprobable parity check codes is

$$M_G = [I_{nR}, M]. \tag{6.32}$$

I_{nR} is an nR by nR identity matrix. M is an nR by $n(1 - R)$ matrix. M's entries are mutually independent binary random variables and each entry has equal probability to be 0 or 1. The codeword to be transmitted is produced by multiplying $\mathbf{t_m}$ with M_G.

$$\mathbf{t} = \mathbf{t_m M_G} \tag{6.33}$$

Here multiplication and addition are bitwise binary operations. The universal set of all codewords is denoted as Ω. The received vector is denoted as $\mathbf{r}$. The receiver uses a parity matrix M_H to check if $\mathbf{r}$ contains bit errors.

$$M_H = \begin{bmatrix} M \\ I_{n(1-R)} \end{bmatrix} \tag{6.34}$$

The parity check process is to multiply $\mathbf{r}$ with M_H. If the result is a 0 vector, the receiver believes that $\mathbf{r}$ does not contain any bit error. The error correction capability of the code is measured by the conditional information of $\mathbf{t}$ given $\mathbf{r}$, i.e. the uncertainty of $\mathbf{t}$ given $\mathbf{r}$. The more uncertainty the receiver has, the less he can do to recover bit errors. The following theorem provides a lower bound for the receiver's uncertainty.

Theorem 6.2 *Suppose the bit error probability of the binary symmetric channel is* p_e*. the receiver's uncertainty of* $\mathbf{t}$ *given* $\mathbf{r}$ *is lower bounded by*

$$H(\mathbf{t}|\mathbf{r}) \geq \mathbf{nh(p_e)} - \mathbf{n(1 - R)h(\phi_{nR+1}(p_e))}, \tag{6.35}$$

with

$$h(x) = -x \log_2 x - (1 - x) \log_2 (1 - x) \tag{6.36}$$

$$\phi_k(x) = \frac{1 + (1 - 2x)^k}{2} \tag{6.37}$$

Proof Please refer to Appendix C.

Equation (6.35) decomposes the lower bound of the receiver's uncertainty into two terms. The first term $nh(p_e)$ is the channel randomness appended to the codeword. The second term $-n(1 - R)h(\phi_{nR+1}(p_e))$ is the redundant information carried in the coded bits which can be used to correct errors. Many practical value sets of n, R, and p_e can ensure a strictly positive lower bound of $H(\mathbf{t}|\mathbf{r})$. Suppose Alice use the equiprobable parity check code to encode a 8192 bits message and append 16 bits checksum to the original message. If the bit error probability is 10^{-3}, $H(\mathbf{t}|\mathbf{r})$ will be greater than 77 bits. In these cases, it is impossible for Eve to recover bit errors by computation.

This lower bound could be negative and provide a theoretical possibility for Eve to recover her bit errors. However, Eve needs to use maximum likelihood (ML) decoding to efficiently utilize the redundant information. ML decoding can be deduced to the coset weights problem, which has been proved to be NP-complete. There is a huge gap between theoretical possibility and error correction in practice.

The error detection capability of the code is measured by the probability of undetected errors, which is defined as $\mathbf{r}$ is believed to contain no error but $\mathbf{t} \neq \mathbf{r}$. The following theorem provides an upper bound on this probability.

Theorem 6.3 *The probability of undetected errors is upper bounded by*

$$p_{ud} = p_{\mathbf{r} \in \Omega}(\mathbf{t} \neq \mathbf{r}) \leq 2^{-n(1-R)} \tag{6.38}$$

Proof Please refer to Appendix C.

Theorem 6.3 states that the probability of undetected errors decrease exponentially with the increase of checksum bits, regardless of the bit error pattern and distribution.

It is possible to design the channel codes that are more sensitive to detect bit errors and less helpful to recover bit errors in more complicated channel models. It is also interesting to consider if there is a fundamental law that governs the trade off between the communication efficiency and the communication security.

Chapter 7
Reliability Analysis for Communication Security

In the previous chapter, we have discussed a lot about information theoretical characteristics of dynamic secrets. In the research community, many researchers would like to categorize dynamic secrets as a heuristic scheme among information theoretic security schemes. However, we would like to emphasize that dynamic secrets are an engineering approach to improve communication security with practical conditions. Furthermore, dynamic secrets provide a different paradigm for communication security system designs.

In this chapter, we examine communication security systems from a reliability engineering perspective. We compare a secure communication system with a machine and the cryptographic key with the key component in the machine. A key theft allows adversary to compromise communication security. Therefore, key thefts are comparable with key component failures which stop the machine from working properly. When a stolen key is replaced by a new key, the compromised secure communication system will reinstate its security status. The machine will return to work once the failed key component is repaired. There is a strong connection between the reliability model of a machine and the security model of a secure communication system when adversary could obtain the key.

The reliability analysis results echo the challenges to the locksmith model presented in Chap. 2 and demonstrate the incompleteness of the long standing Kerckhoffs principle. We propose to evaluate *reliability of communication security* for secure communication systems and improve their designs.

Firstly, we demonstrate that a common security practice, "change password once three months", provide little help to protect users' accounts from password thefts. In fact, all the periodic key update schemes have an inherent limitation that prevent them from promptly replacing the stolen key. The analysis suggests that the existing digital certificate system, the digital ID system of the Internet, might be unreliable and provides a false feeling of security in a significant portion of time. On the other hand, reliability analysis finds that dynamic key updates could be an effective countermeasure to key thefts and improve the availability of communication security.

S. Xiao et al., *Dynamic Secrets in Communication Security*,
DOI: 10.1007/978-1-4614-7831-7_7,
© Springer Science+Business Media New York 2014

Secondly, the typical design of two-factor authentication systems that uses an electronic security token and a password in authentication cannot provide security in a consistent manner. Reliability analysis shows that this typical system design aims to delay the first time of security breach instead of maximizing the availability of security and the speed that the system recovers from key thefts. We could use two dynamic keys to construct a two-factor authentication system and achieve superior security performance than this typical design.

Section 7.1 introduces basic ideas of the reliability engineering theory with an engineering system example. Section 7.2 formulates the key safety problem as an engineering reliability problem, and then relates the engineering reliability concepts to communication security. Section 7.3 compares traditional periodic key update schemes and the dynamic key update scheme using the reliability measures. The analytic results reveal a fundamental weakness in traditional key management, which is used for daily online account password management and the management of digital certificates across the Internet. The results also suggest that the dynamic key update scheme is a viable enhancement to existing secure communication systems. Section 7.4 extends the reliability analysis to two-factor authentication systems that are trusted and widely used in security sensitive domains such as online banking systems and military networks. The analytic results demonstrate factors that significantly affect the security performance of these systems.

7.1 Reliability Theory Preliminaries

We introduce reliability concepts by considering a simple engineering system with only one component. The system is working if the component is in good status and malfunctioning if the component is in a failed state. The failed component can be repaired and return to work. A canonical example of this one component system is a lamp with a light bulb that may burn out after being used for a random period of time. The failed light bulb can be replaced and then the lamp illuminates again.

A two-state reliability model is used to describe this system, as shown in Fig. 7.1. The system is working at time 0. After a period of time, the component experiences a failure and the system enters the failed state. Later, the system undergoes maintenance and returns to the working state. The system state transitions over time are demonstrated as an ON-OFF process.

This model is easily understood in the context of the lamp example. The lamp is working at the beginning and keeps illuminating for a period of time after which its bulb burns out and the light goes off. After the failure is detected and the bulb replaced, the lamp will continue to illuminate.

To quantitatively study the reliability properties of this system, we define $X(t)$ as the system state variable.

$$X(t) = \begin{cases} 1, & \text{if system is working} \\ 0, & \text{if system is failed} \end{cases} . \tag{7.1}$$

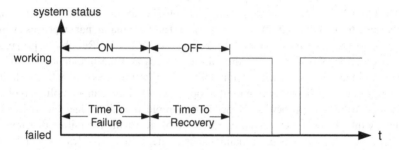

Fig. 7.1 An ON-OFF process model for the reliability analysis of an one component engineering system

The first metric that we consider is the **Mean Time To Failure (MTTF)**. MTTF is the expected time that the system works without failure. Let $N_f(T)$ denote the number of failures in $[0, T]$. MTTF can be expressed as

$$MTTF = \lim_{T \to \infty} \frac{\int_0^T X(t)dt}{N_f(T)}. \tag{7.2}$$

In the lamp example, MTTF corresponds to the average time that a bulb illuminates before burning itself.

A second metric of interest is the **Mean Time To Recover (MTTR)**, the average time it takes to recover the system from a failure. MTTF is defined as

$$MTTR = \lim_{T \to \infty} \frac{\int_0^T (1 - X(t))dt}{N_f(T)}. \tag{7.3}$$

MTTR for the lamp example is the mean time from the moment when the bulb burns out till someone notices the burned out bulb and replaces it with a new one.

Average availability is a widely used metric. It characterizes the fraction of time that the system functions properly over the total system running time. The average availability is defined by integrating $X(t)$ over time,

$$A_{avg} = \lim_{T \to \infty} \frac{\int_0^T X(t)dt}{T}. \tag{7.4}$$

It can also be derived from MTTF and MTTR.

$$A_{avg} = \frac{MTTF}{MTTF + MTTR} \tag{7.5}$$

In the lamp example, the average availability is the portion of time that the light is on over the total time that the lamp is in use.

The ON-OFF process can be decomposed into the failure process which contains all the state transitions of $X(t) = 1 \rightarrow X(t) = 0$ and the maintenance process which contains all the state transitions of $X(t) = 0 \rightarrow X(t) = 1$. Reliability properties are governed by the statistics of the failure process and the maintenance process. For an engineering system with certain failure distribution, i.e. distribution of failure process, reliability theory can be used to optimize the maintenance policy to obtain the desired reliability properties. In the lamp example, the failure distribution is determined by the physical attributes of the light bulbs. Knowing the failure distribution, the lamp operator can design a maintenance strategy that provides a small MTTR and large average availability with a minimal effort to check the light and replace the bulb.

Many other metrics can be used to characterize an engineering system than the MTTF, MTTR, and average availability. In this chapter, we focus on these three properties to analyze the reliability of communication security.

7.2 Key Safety as a Reliability Problem

In this section, we consider a secure communication system which is only vulnerable to key theft. This system can be modeled as a one component reliability engineering system where the cryptographic key is the only component. Key safety determines communication security.

As shown in Fig. 7.2, $X(t) = 1$ denotes that the communication system is secure at time t. $X(t) = 0$ means communication security is compromised. Key theft is a system failure that results in a transition from $X(t) = 1$ to $X(t) = 0$. When the stolen key is replaced by a new key, communication security recovers.

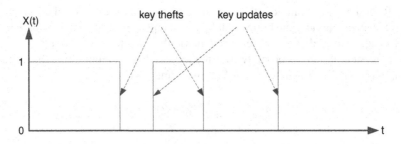

Fig. 7.2 An ON-OFF process model for the reliability of the key safety

7.2.1 Reliability Properties in Communication Security Context

The engineering system reliability metrics take on new meanings in the communication security context.

MTTF is the average time that a cryptographic key remains secret before it is obtained by the adversary. This metric is determined by the key protection techniques utilized in the system and the dedication and technical strength of the adversary. A dedicated adversary equipped with advanced offensive technologies is expected to crack or steal the key in less time than a casual adversary who is only motivated by curiosity.

MTTR is the average time required for the communication system to restore its security. It is also the average length of a time interval during which the adversary can exploit the stolen key. If a secure communication system is characterized by a very small MTTR, the adversary's incentive to attack is considerably weakened.

The average availability demonstrates the fraction of time for which the communication system is secure. We can also define **average unavailability** accordingly,

$$U_{avg} = 1 - A_{avg} \tag{7.6}$$

The reliability of communication security is characterized by the MTTF, MTTR, and average availability of the key.

We decompose the ON-OFF process of communication security into two interleaved processes. One process characterizes all key thefts and corresponds to the failure process of an engineering system. A second process contains all key updates that correspond to preventive maintenances and repairs for an engineering system. In practice, users generally have no control over the time when the adversary obtains the key. In order to optimize the reliability measures for communication security, users should carefully choose a key update scheme which determines the key update process.

7.2.2 Poisson Process Modeling of Key Thefts

In order to analyze reliability, we need a model for the key theft process. There is no straightforward data that can be used to characterize this process because the adversary usually does not notify the users when he obtains the key. Users are only aware of the key theft after noticeable damage have been caused.

Consider a sequence of key theft events occurring over time in a secure communication system. The timing of these key thefts can be affected by many independent or weakly correlated factors such as vulnerabilities in the victim system, network connectivity for the key stealing attack, the computing power of adversary, the adversaries' attack timing preferences, user behavior patterns, and numerous other constraints. It is reasonable for a first order analysis to assume that the sequence of key theft events form a Poisson process. In our case, we argue that the perception of unpredictability for the timing of the next key theft seems to support a memoryless inter key theft time distribution. The rate of key theft events is time invariant and denoted as λ. References [55] and [56] provide some experimental results to support the time invariant Poisson approximation of security incidents.

Another approach that results in the Poisson process model is to explore the "no aging" property of the key. A cryptographic key is a string of digits that does not wear out over time. This implies an exponential life distribution. There is no reason to argue that one key can be more easily stolen than any other key. Therefore the exponential life distributions of the keys are identical. We further assume that key thefts are independent, i.e. previous key thefts do not make the future key thefts easier. Then the key theft process would be a time invariant Poisson process.

7.3 Reliability Analysis for Key Update Schemes

This section compares the most widely adopted key update scheme, the periodic key update scheme, with the dynamic key update scheme presented in Chap. 3 using the above defined reliability measures.

7.3.1 Periodic Key Update

Periodic key updates are quite common in practice. A notable example is the IT security advice that prompts a user to change the online account password once every three months. The digital certificates used to authenticate the legitimacy of Internet websites follow the periodic key update scheme as well. After a period of time, typically a year, the public key signature for the digital certificate would expire and the user needs to renew the digital certificate.

If we assume that key thefts occur over the time, then the key safety can be characterized by an ON-OFF process, as shown in Fig. 7.3.

The key update period is denoted as T. In practice, the value of T is much smaller than the average time between two successive key theft events, $1/\lambda$. We may assume at most one key theft occurs between two successive key updates. We can calculate MTTR of periodic key update scheme using the memoryless property of the exponential distribution.

$$
\begin{aligned}
MTTR_P &= \sum_{k=1}^{\infty} \int_{(k-1)T}^{kT} (kT - t)\lambda e^{-\lambda t} dt \\
&= \frac{T}{1-e^{-\lambda T}} - \frac{1}{\lambda}
\end{aligned}
\tag{7.7}
$$

It follows from Eq. (7.7) that $MTTR_P$ is lower bounded by

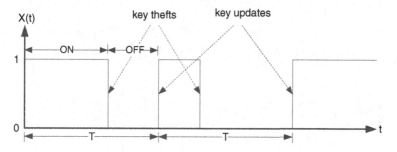

Fig. 7.3 An ON-OFF process model for the reliability of the periodic key update scheme

$$MTTR_P > \frac{T}{2} \qquad (7.8)$$

for all $T > 0$ and $1/\lambda > 0$.

Inequality (7.8) uncovers a fundamental inefficiency of the periodic key update scheme. **The average time to recover a stolen key is always greater than half of the key update period**. If we expect a stolen cryptographic key to be replaced within 24 h, then the key update period T has to be less than 48 h. Such a relentless key update scheme is impractical for most real world applications. On the other hand, if T is a large, practical value, such as $T = 3$ months, then the adversary will have more than one month to attack with the stolen key. The periodic key update fails its purpose to promptly replace the stolen key.

Empirical studies such as [49] and [39] argue that the periodic password change policy is ineffective and cannot justify its cost. The reliability analysis demonstrates a theoretical rational that support their observations.

We further examine the average availability of a cryptographic key with the periodic key update scheme,

$$A_{avg,P} = \frac{1/\lambda}{1/\lambda + MTTR_P} = \frac{1}{\lambda T}(1 - e^{-\lambda T}) \qquad (7.9)$$

$$U_{avg,P} = 1 - A_{avg,P} = 1 - \frac{1}{\lambda T}(1 - e^{-\lambda T}) \qquad (7.10)$$

Consider an example of a three-month periodic password update scheme for a user's online account. Suppose on average the adversary takes 1 year to crack the user's password, i.e. $1/\lambda = 12$ months. The value of $U_{avg,P}$ is more than 0.115, which indicates that the user would expect his online account to be unprotected for approximately one month every year. If the password is stronger and it takes the adversary 10 years on average to crack it, we still have $U_{avg,P} > 0.012$. Such a reliability of communication security is still far from satisfactory for any security sensitive task.

Figure 7.4 plots the average availability of the periodic key update scheme with various T and $1/\lambda$.

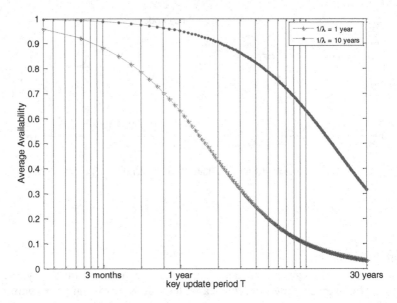

Fig. 7.4 Average availability of the periodic key update scheme for $1/\lambda = 1$ year and $1/\lambda = 10$ years. Typical key update periods are noted on X-axis. 3 months corresponds to the regular web password updates. 1 year is the typical expiration period for a personal digital certificate. 30 years is the life span of the root certificate used by Internet certificate authorities

Consider the digital certificate system used in the Internet. There are hundreds of root certificates stored in our Internet browsers. These root certificates are the roots of the trust hierarchy in the Internet. The private keys of these root certificates are stored in top level CA companies and protected with extreme caution. The safety of these keys are critical to the Internet security. If adversary obtains any of these keys, he can create fake websites that mimic legitimate sites such as www.google. com. Our Internet browser would be deceived because the fake websites may. The root certificates generally have very long life span, typically several decades.

From Fig. 7.4, we can estimate the safety requirement for the private keys of the Internet root certificates. The curve suggests that if any adversary can obtain a root private key in 10 years, our current Internet digital certificate system will not be dependable because the average availability of the root certificates, i.e. the portion of time that a root certificate is trustworthy, is below 0.4. The top tier certificate authority (CA) needs to guarantee that the key theft rate is less than once a hundred years or even smaller to provide a reasonable average availability for the Internet digital certificate system.

7.3.2 Dynamic Key Update

Assume the secure communication users choose the dynamic key update scheme from Sect. 3.2 for cryptographic key management. Recall the packet communication model of dynamic secrets presented in Chap. 3. The dynamic key update scheme is defined by Eqs. (3.1) and (3.2).

In order to analyze the reliability of this scheme, the following notations are introduced to the original dynamic secrets model. The communication environment is noisy and the packet loss probability is at least p_l, regardless of the receiver. Alice and Bob exchange packets at a constant rate R.

Alice and Bob generate a dynamic secret s_i for every delivered packet p_i using a hash function $f_H(\cdot)$,

$$s_i = f_H(p_i). \tag{7.11}$$

Then they both update the cryptographic key,

$$k_i = k_{i-1} \oplus s_i \tag{7.12}$$

k_i is used to secure the transmission of the next packet p_{i+1}. Such a dynamic key update scheme changes the key value every time a packet is delivered either from Alice to Bob or from Bob to Alice.

The adversary, Eve, has the capability of obtaining the cryptographic key at some points in time. She is also allowed to eavesdrop the Alice-Bob communication without being caught. Her eavesdropping will not be perfect because of the noise in the communication environment. Eve is assumed to be equipped with the most advanced receiving technology, which allows her packet loss rate to approach the lower bound of p_l in this communication environment.

The reliability model for the dynamic key update scheme is a two-state Markov model and is shown as Fig. 7.5.

The dynamic key update scheme cannot prevent key theft. Therefore the transition rate from state G to state B corresponds to the key theft rate, λ.

After every key theft, the adversary is forced to keep eavesdropping and track the key updates because the key is updated with every packet. Due to environmental noises, packet loss will inevitably occur in the adversary's eavesdropping and the adversary will not be able to know the updated key. Communication security recovers

Fig. 7.5 The state transition model for the dynamic key update scheme. G state means the cryptographic key is a secret. B state means the key is known by the adversary

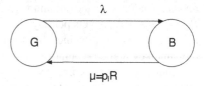

as soon as the adversary encounters a packet loss. Therefore the transition rate from B to G is the packet loss rate for the adversary, $\mu = p_l R$.

We can calculate the MTTF, MTTR, and average availability for a cryptographic key when the dynamic key update scheme is applied; they are

$$MTTF_D = \frac{1}{\lambda} \tag{7.13}$$

$$MTTR_D = \frac{1}{\mu} = \frac{1}{p_l R}. \tag{7.14}$$

$$A_{avg,D} = \frac{1/\lambda}{1/\lambda + 1/\mu} = \frac{p_l R}{\lambda + p_l R}. \tag{7.15}$$

$$U_{avg,D} = 1 - A_{avg,D} = \frac{1/\mu}{1/\lambda + 1/\mu} = \frac{\lambda}{\lambda + p_l R}. \tag{7.16}$$

Unlike the periodic key update case, the MTTR for dynamic key update scheme can be effectively decreased by exchanging packets at a higher rate and introducing more noise into the communication environment so as to increase the adversary's loss probability. When the MTTR decreases, the average availability increases.

We plot the MTTR and the average availability as functions of $p_l R$ in Fig. 7.6. The average availability is also affected by the value of $1/\lambda$. Figure 7.6 contains three average availability curves corresponding to three different values of $1/\lambda$. As $p_l R$ increases, the MTTR of the dynamic key update scheme inverse-proportionally decreases, and the average availability quickly approaches one.

We can check the MTTR and the average availability for a secure WiFi connection that utilizes the dynamic key update scheme, such as the experimental office wireless LAN (802.11g) described in Chap. 4. In this scenario, the adversary's packet loss rate, p_l, is influenced by many factors such as packet communication rate, packet size, eavesdropping distance, environmental background noise level, and the signal strength of the interference generated by neighbor office wireless LANs. Because it is an office environment, the adversary may not be able to deploy dedicated eavesdropping hardware such as a huge gain-boost antenna. He may use, instead, a commercial-off-the-shelf (COTS) WiFi receiver to eavesdrop the users' wireless communications. A typical packet loss rate of such an adversary would be $p_l = 10^{-4}$ [65].

The users have control over their packet communication rate R. In order to make a conservative estimate, we assume the two users only use a small portion of the available bandwidth and communicate at $R = 10$ packets per second, or equivalently, $R = 8.64 \times 10^5$ packets per day. We are quite cautious in estimating both p_l and R. $p_l R = 86.4$ is easily achievable for two users in a wireless LAN that utilizes the dynamic key update scheme.

From Fig. 7.6, the corresponding MTTR is 0.01 day. A successful key stealing attempt is expected to grant the adversary only about 15 min time to attack the secure wireless communication. The adversary will likely be discouraged from stealing the

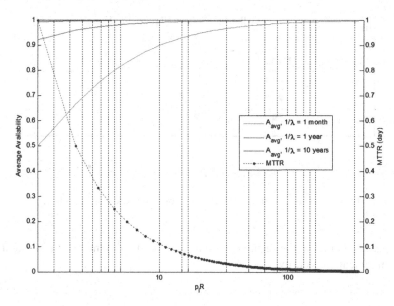

Fig. 7.6 Average availability of dynamic key updates for $1/\lambda = 1$ month, $1/\lambda = 1$ year, and $1/\lambda = 10$ years. MTTR as a function of $p_l R$ value is also plotted. The typical value of $p_l R$ for an office wireless LAN is noted on X-axis

key since the expected gain is severely limited. From the user's point of view, the stolen key is expected to automatically recover within 15 min. The recovery can be faster if the users communicate at a higher rate.

The average availability shows the fraction of time for which the communication is secure. If the key is frequently obtained by the adversary, the average availability will be low. Normally, the secure communication system is designed to provide reasonable protection to its key such that the adversary needs to spend an unrealistic amount of time and effort to crack and obtain the key. However, key thefts will be frequent when a serious cipher vulnerability is found by the adversary or the advancement of computing technology allows the adversary to crack a short key by brute force [23, 46].

Consider a secure communication system that has a design vulnerability that allows the adversary to obtain the cryptographic key once a month on average ($1/\lambda =$ one month). Such a system is not usable in practice. Figure 7.6 shows, even in this extreme case, that the dynamic key updates still maintain a certain level of security protection by providing an average availability of communication security greater than 0.9996.

In many practical scenarios, the dynamic key update scheme is capable of providing a smaller MTTR and a higher average availability than the periodic key update scheme, as shown in Fig. 7.7. However, when the users rarely communicate with each other or the communication environment is less error-prone, the reliability performance of the dynamic key update scheme degrades. We can combine and use the two

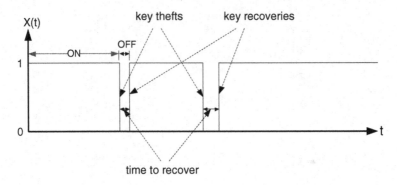

Fig. 7.7 An ON-OFF process model for the reliability of dynamic key update scheme

key update schemes in the secure communication system to improve the reliability of communication security.

7.3.3 Session Keys and Key Reliability

Prof added—One approach that has been proposed to reduce the time for which a key is vulnerable is... Session key exchanges allow...

It is worth noting the difference between key updates and session key exchanges. Session keys are used to defend against cryptanalysis attacks by limiting the amount of data processed using a particular key value. Session key exchanges allow users to frequently change encryption keys within a communication session. As shown in Fig. 7.8, the master key is used to initiate every secure communication session and a session key is only used during its corresponding session.

All session keys are derived from the master key through a session key generation protocol. The secrecy of all session keys relies on the secrecy of the master key. Session key exchanges do not change the value of the master key. Therefore, frequently exchanging session keys does not affect the reliability properties of the master key and, consequently, has no influence on the reliability of communication security.

Fig. 7.8 The master key and session keys. k_m is the master key. k_{xy} with $x = 1, 2, y = 1, 2, \ldots$ are session keys

7.4 Reliability Analysis for Two-Factor Authentication Schemes

The analysis in the previous section reveals the unreliability of passwords under periodic key management. In practice, people also find that the password authentication is not sufficiently trustworthy to safeguard security sensitive accounts. Therefore, two-factor authentication scheme is proposed to provide stronger authentication. The two-factor authentication scheme requires the user to submit the password and one additional credential together as the proof of identity.

This section focuses on the reliability analysis of two-factor authentication schemes and compares an authentication scheme that uses an electronic token and a password with an authentication scheme that uses two independent dynamic keys. The analytic results not only demonstrates the difference in the reliability performances of these two schemes, but also reveal the difference in the design objectives.

7.4.1 Electronic Token-Password Authentication Scheme

The electronic token-password authentication scheme is a typical example of a two-factor authentication scheme. The user remembers a password and possesses an electronic security token. The security token is programmed to display a pseudo random code that changes frequently, typically once every 30 s. When a user attempts to login, he submits his password together with the pseudo random code displayed on his token at the time of login. The pseudo random codes are generated from a seed value. The authentication server stores the seed value therefore it can verify if the submitted pseudo code is from the user's token by synchronously generating the same pseudo random code.

The seed value and the pseudo random code generation algorithm must be carefully protected. Once an adversary knows them, he can calculate the pseudo random codes for any time. On the other hand, the seed value and the generation algorithm will remain unchanged for a very long period of time because any change to them will incur a heavy cost to upgrade and redistribute electronic security tokens to the users. In practice, security tokens are often not upgraded until there is firm evidence to prove that the adversary can reproduce the pseudo random codes.

The electronic token-password authentication scheme can be characterized by a two-component reliability model [13], shown as Fig. 7.9. The password and the electronic security token are two components in the system. The authentication scheme fails when both components fail, i.e. an adversary obtains both the password and the seed value.

The password and the electronic security token have distinctive reliability properties. The password is susceptible to various password stealing and cracking attacks. The failure rate of the password component can be substantial in practice. On the other hand, the password can be periodically updated at an affordable cost. A password theft will not cause the permanent failure of the password component.

Fig. 7.9 The combination
of electronic security token
and password improves the
reliability of authentication

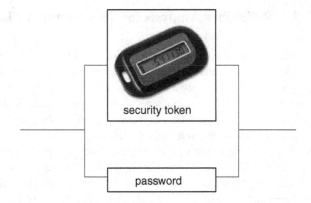

The security token component has a very low failure rate. It is extremely difficult to penetrate the heavily protected authentication server to obtain the seed value. On the other hand, updating the seed value would require re-programming the token hardware. Such an update is complex and expensive. As a result, the seed value practically never changes. If an adversary obtains the seed value, the electronic token component may remain in the failed state for a long period of time.

We apply the ON-OFF modeling technique to characterize the electronic token-password authentication scheme, as shown in Fig. 7.10. The ON-OFF process for the authentication scheme is the superposition of the ON-OFF processes of the two components.

In the time period before the electronic security token component fails, the authentication result is always trustworthy regardless of the password safety status. After the failure of the electronic token component, the safety of password component would decide the dependability of the authentication outcome. If the failure rate of

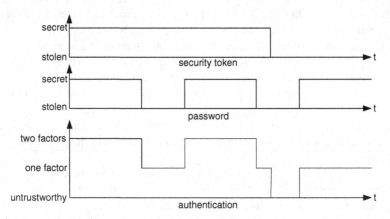

Fig. 7.10 ON-OFF process model for the two-factor authentication scheme using security token and password

the token component is not zero, i.e. it will fail at some time point, MTTF, MTTR, and average availability of the electronic token-password authentication scheme will be the same as those of the password only authentication scheme.

$$MTTF_{token\text{-}password} = MTTF_{password} \tag{7.17}$$

$$MTTR_{token\text{-}password} = MTTR_{password} \tag{7.18}$$

$$A_{avg,token\text{-}password} = A_{avg,password} \tag{7.19}$$

The electronic security token only helps the authentication scheme prior to its first failure. It improves the scheme's reliability in the transient period. Specifically, the use of the electronic security token prolongs the mean time to first system failure (MTFSF) of the two-factor authentication scheme.

MTFSF is the average amount of time that a system keeps working before the system fails for the very first time. For an electronic token-password authentication system, its MTFSF is the average time length before the first time both the token and the password fail. We have

$$MTFSF_{token\text{-}password} \geq \max\{MTFSF_{token}, MTFSF_{password}\}. \tag{7.20}$$

Because the token has a much lower failure rate than the password, in practice,

$$MTFSF_{token} \gg MTFSF_{password} \tag{7.21}$$

and

$$MTFSF_{token\text{-}password} \approx MTFSF_{token} \tag{7.22}$$

only prior to the failure of the electronic token. The token-password scheme is more reliable than the single password scheme. Therefore the electronic token-password system design has to optimize the protection to the seed value in order to increase the $MTFSF_{token}$. As long as the token is secure, the authentication result is trustworthy.

This design philosophy is challenged by real world security incidents. One prominent incident is the seed value theft of SecureID™. SecureID™ is one of the most widely deployed security token system in the world. It is deployed in admission-critical places such as Pentagon and Lockheed Martin Company. The seed value of SecureID™ was found to be leaked to adversarial groups in May, 2012 [64]. Later, its pseudo codes generation algorithm was also reported cracked [11]. In practice, the MTFSF of the electronic token-password authentication scheme may be much lower than its design expectation.

7.4.2 Authentication with Two Dynamic Keys

It is possible to construct a two-factor authentication scheme with two independent dynamic keys. For example, a user can use both his computer and mobile phone to communicate with the authentication server. His computer is used to login to the server and perform various operations with the online account. The computer-server communication produces one dynamic key, denoted as k_c. His mobile phone only sends one randomly generated packet every 30 s to the authentication server and updates another dynamic key, denoted as k_m.

When the users attempts to login, he uses the computer to connect to the authentication server and inputs the hash value of k_m which can be automatically displayed on his mobile phone. From the user's perspective, the two-dynamic key authentication scheme is more convenient than the electronic token-password authentication scheme because the user no longer needs to memorize and manage the password. The reliability model of the two-dynamic key authentication system is shown in Fig. 7.11.

As discussed in the previous section, the safety of each dynamic key can be modeled by a two-state Markov model. Therefore, the reliability of the two-dynamic key authentication system can be modeled as a four-state Markov model, shown in Fig. 7.12.

In the figure, λ_m and λ_c are the failure rates of k_m and k_c. $\mu_m = p_{l,m} R_m$ is the recovery rate of k_m in case its stolen. μ_m equals to the product of the mobile phone's packet exchange rate R_m and the packet loss rate $p_{l,m}$ when an adversary is eavesdropping the communication between the mobile phone and the authentication server. μ_c is defined similarly as μ_m.

We can find analytical solutions for the MTTR, average availability, and MTFSF for this model. The average time to escape from state B, i.e. the average time to recover communicate security from a breach, is

Fig. 7.11 The combination of two dynamic keys form a reliable two-factor authentication scheme

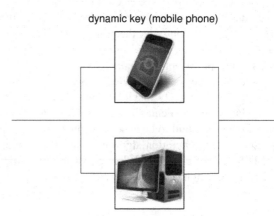

dynamic key (mobile phone)

dynamic key (computer)

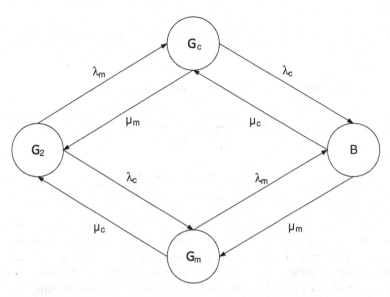

Fig. 7.12 Four-state Markov model for dynamic key based two-factor authentication scheme. G_2 state means both keys are secret. G_c state means only the dynamic key on computer is secret. G_m state means only the dynamic key on mobile phone is secret. B state means both keys are known by adversary and the communication security is lost

$$MTTR_{two\text{-}key} = \frac{1}{\mu_c + \mu_m} \tag{7.23}$$

The average availability of this authentication scheme is

$$A_{avg,two\text{-}key} = 1 - \frac{1}{2\left(1 + \frac{\mu_c\mu_m}{\mu_c + \mu_m}\frac{\lambda_c + \lambda_m}{\lambda_c\lambda_m}\right)} \tag{7.24}$$

The mean time to the first system failure is

$$MTFSF_{two\text{-}key} = \frac{1 + \frac{\lambda_m}{\mu_m + \lambda_c} + \frac{\lambda_c}{\mu_c + \lambda_m}}{\lambda_c\lambda_m\left(\frac{1}{\mu_m + \lambda_c} + \frac{1}{\mu_c + \lambda_m}\right)} \tag{7.25}$$

The detailed derivation of the $MTFSF_{two\text{-}key}$ is found in Appendix D.

We introduce the following intermediate variables to simplify the results in Eqs. (7.24) and (7.25),

$$\lambda = \max\{\lambda_c, \lambda_m\} \tag{7.26}$$

$$\mu = \min\{\mu_c, \mu_m\} \tag{7.27}$$

$$\alpha_c = \mu_c/\lambda \tag{7.28}$$

$$\alpha_m = \mu_m / \lambda \qquad (7.29)$$

$$\alpha = \mu / \lambda \qquad (7.30)$$

Then the average availability and MTFSF can be interpreted as

$$A_{avg,two\text{-}key} \geq 1 - \frac{1}{2\left(1 + \frac{2\alpha_c \alpha_m}{\alpha_c + \alpha_m}\right)} \geq 1 - \frac{1}{2(1+\alpha)} \qquad (7.31)$$

$$MTFSF_{two\text{-}key} \geq \frac{1}{\lambda}\left(1 + \frac{(1+\alpha_c)(1+\alpha_m)}{2 + \alpha_c + \alpha_m}\right) \geq \frac{1}{2\lambda}(3+\alpha) \qquad (7.32)$$

If the user carefully protects both k_m and k_c, the corresponding key theft rates, λ_m and λ_c, will be small. The user's mobility will result in substantial packet losses to the adversary and provide a high recovery rate for k_m, i.e. μ_m is large. The user can maintain a high communication rate between his computer and the authentication server to obtain a large μ_c. In many practical scenarios, the user will have

$$\min\{\mu_c, \mu_m\} \gg \max\{\lambda_c, \lambda_m\} \qquad (7.33)$$

and

$$\alpha \gg 1. \qquad (7.34)$$

Subsequently, $A_{avg,two\text{-}key}$ is close to one and $MTFSF_{two\text{-}key}$ can be very large.

Compared with the electronic token-password authentication system, whose MTTR is the same as the MTTR of the single password authentication system, the two-dynamic key authentication system improves its MTTR over the single dynamic key system.

$$MTTR_{two\text{-}key} = \frac{1}{\mu_c + \mu_m} < \min\left\{\frac{1}{\mu_c}, \frac{1}{\mu_m}\right\} \qquad (7.35)$$

A more fundamental difference is on the design to achieve a large MTFSF. The electronic token-password authentication system employs one component with a small failure rate and the system's MTFSF approximately equals the MTFSF of this component. The two-dynamic key authentication system uses two components that each has a small MTTR. Each component may occasionally fail. Because the MTTR is small, the failed component quickly recovers and the probability that both components fail at the same time is small. Subsequently, the MTFSF is large.

The MTFSF design in the electronic token-password authentication system is limited by practical factors such as the cost constraint. It is difficult yet inefficient to have a component with extremely high MTFSF in order to have a high MTFSF for the system. In comparison, the two-dynamic key authentication system can be naturally extended to include more components. Each additional component reduces the probability that all components fail at the same time and improves the MTFSF.

Chapter 8
Future Applications

In previous chapters, we presented dynamic secrets and identified security features that complement and improve traditional secure communication systems. In this final chapter, we will present some application ideas that can possibly make existing secure communication applications more efficient, easier to use, and more resilient to cryptographic key-related attacks. It would be our great pleasure if readers find more ways of using dynamic secrets.

In Sect. 8.1 we propose the use of dynamic secrets to secure business transactions for mobile users. Dynamic secrets may replace traditional usernames and passwords to build a strong and unique credential to prove the legitimate identity of users. In Sect. 8.2 we expand the idea of dynamic secrets to the software copyright regime. The use of dynamic secrets can help software makers distinguish a legitimate software copy installed on multiple computers from a pirate copy. Section 8.3 discusses the potential collaboration between dynamic secrets and energy efficient wireless communication schemes. The reduction on transmission power could be a positive factor for communication security because the eavesdropping adversary would have higher information loss rate.

8.1 Secure Mobile Transactions

The increasing accessibility and bandwidth of mobile communications provide a booming mobile business market in which users carry out business transactions on their mobile computing devices, such as smart phones and tablets [38]. Security is of critical importance in this case. Users would hesitate to use a mobile business platform if they are not sure about the reliability, authenticity, and confidentiality of their business transactions.

Authentication is an essential component to ensure the security of mobile transactions. A user would demand the guarantee that only he himself can access his account and complete a mobile transaction. It is difficult to provide such a guarantee [75].

S. Xiao et al., *Dynamic Secrets in Communication Security*,
DOI: 10.1007/978-1-4614-7831-7_8,
© Springer Science+Business Media New York 2014

The most typical method to verify a user's identity for mobile business transactions is the username-password authentication scheme, which is also widely used for Internet online transactions. As discussed in Chap. 2, password-based authentication is inherently vulnerable to exhaustive search attack because our brains are not good at remembering high entropy passwords. The vulnerability becomes worse for mobile transactions. It is troublesome for users to input long and random passwords on a mobile device [42].

Various schemes are proposed to allow users to be authenticated on a mobile device without inputting passwords. A simple upgrade is to ask users to store signed digital certificates on their mobile devices to replace passwords as authentication credentials [2]. This approach is less popular because it demands users to have security expertise to handle digital certificates. It also requires mobile business platforms to maintain complex and expensive public key infrastructures. Some propose taking users' behavior patterns as authentication credentials, as presented in Refs. [109] and [57]. These authentication schemes are limited by the inaccuracy of the statistics on users' macroscopic behaviors.

Dynamic secrets may provide an alternative approach for user authentication in mobile transactions. When a user opens his account on a mobile business platform, he registers his mobile devices by establishing a dynamic key between each mobile device and the platform. Communications between his mobile device and the platform updates the key constantly. It is impossible for an adversary to keep tracking key updates because user mobility and wireless randomness.

Moreover, dynamic secrets provide an inherent detection method to impersonation attacks. Even if an adversary clones a user's smart phone and use it to connect to the platform, the user will surely detect the existence of the clone version of his smart phone. The mobile business platform can further strengthen its security measures by requiring users' mobile devices to maintain persistent communication with the platform at a low speed and delaying every business transaction by a short amount of time.

Suppose a user's smart phone sends a beacon packet to the platform every 10 s. These beacon packets are also used to generate dynamic secrets. The platform delays every transaction request by 30 s before executing it. A fraudulent transaction request sent from a cloned smart phone will be detected before getting executed. The sending frequency of beacon packets could be adjusted according to the energy and bandwidth constraints of the mobile device. The transaction delay should be jointly decided by the amount of money involved in the transaction and the user's tolerance to the delay.

From a user's point of view, mobile transactions that rely on dynamic secrets are more user-friendly because the user does not need to input a password nor install a digital certificate. The transactions are guaranteed to be authentic as long as the user can control the access of his mobile devices.

The same idea can be applied to other scenarios in which users demand communication security for their wireless devices. For example, doctors may carry medical devices with wireless connections and travel around the hospital to see patients. Dynamic secrets could be used to provide a user-friendly solution to secure the data transfer for medical devices.

8.2 Software Identity by Patching History

Software piracy is a serious concern for the software industry. Pirated copies of a software cause substantial financial losses to software makers and can serve as carriers of malware, which threatens the information security of users. It is difficult to prevent software piracy because, if a legitimate software works properly, an exact copy of this software should work as well.

When a software is a stand-alone product, it is fundamentally difficult to prevent it from being pirated [90]. If a stand-alone software works properly, a bit-by-bit copy should work as well in an identical environment. The ultimate anti-piracy technique for stand-alone software involves hardware, if the software does not detect the existence of its corresponding hardware, the software stops working. This anti-piracy approach is expensive. It can be defeated by either removing the code of checking hardware from a legitimate software before making copies, or duplicating the hardware.

It is possible to implement a software that only works when it is connected to an application server. As a result, the application server can count the number of active instances of the software and effectively restrict pirate usages. However, this approach limits the software usage as well. Even with a legitimate software, a user can only use it when the application server is present. Moreover, maintaining a constantly available and accessible application server is expensive. The pirate adversary may defeat the application server based approach by duplicating the application server together with the software [81].

In practice, the software industry depends more on social and economic approaches to mitigate the software piracy problem [78, 131]. Software makers put watermarks into genuine copies of the software [136]. Later software makers examine the pirated copies and pursue the owner of the corresponding genuine copy. The pirate adversary may distort the watermark in the pirated copies to confuse software makers.

In the Internet era, software makers may require users to register their genuine copies online during installation. Software makers then record the computer hardware information where the software is installed. Subsequently, often when a software patch is issued, a software maker checks the genuineness of the copies of its software. if the software maker finds that a software copy works on a computer that consists of different hardware from the software copy's installation record, this copy is marked as a pirated copy. A famous example of this approach is the Microsoft Windows Genuine Advantage (WGA) system [96]. The downside of this approach is, it will incorrectly punish a legitimate user who upgrades his computer hardware and let go of a pirated copy that runs on a virtual machine simulating the hardware environment in which a genuine copy is installed.

Recall the idea of using dynamic secrets to detect a duplicate key in Chap. 3. It is possible to use dynamic secrets to distinguish a pirated copy from a genuine copy even if they are identical. We propose generating dynamic secrets for a software copy from its history of the applied software patches.

Suppose a genuine copy of a software automatically downloads and applies software patches. Let $p_1, p_2, \ldots, p_n$ denote the software patches downloaded at time $t_1, t_2, \ldots, t_n$. At time t_i, the software maker's patch server and the genuine software copy calculate a dynamic secret,

$$s_i = f_H(p_i\|t_i) \tag{8.1}$$

for $i = 1, 2, \ldots$. For $t > t_1$, both the server and the software copy store the dynamic key

$$k(t) = \bigoplus_{i, t_i < t} s_i \tag{8.2}$$

At time t_{n+1}, when a new patch is available and the genuine copy approaches the server to request a download, $k(t_n)$ serves as a proof of genuineness.

Because $k(t)$ is not hardware-related, the genuine software user can change its computer hardware as he wishes. On the other hand, the pirated copies cannot get any software patches without $k(t)$. If a pirated copy goes with $k(t_n)$ and updates itself, the genuine copy will no longer be eligible for software patches. The genuine software user will then report to the software maker and uncover the existence of piracy.

In order to force pirated copies to apply patches and expose themselves, the software could implement usage counters in its program such that the software will crash after a certain number of usages. The counters are reset whenever a patch is applied.

8.3 Energy Efficient Secure Wireless Communications

The reduction of energy consumption in wireless transmissions is an important topic in wireless communication research. Besides the research for advanced materials and devices, researchers proposed improving energy efficiency by adjusting transmission strategy. For examples, MIMO transceivers can save energy through cooperative transmissions [115]; wireless link layer protocols can be modified to be more energy-efficient [36]; in a wireless network, energy consumption can be reduced through routing protocol design and cross-layer optimization [79, 88].

Dynamic secrets can provide a stronger protection to communication security when the adversaries are more likely to lose information in eavesdropping. Decreasing the energy used in wireless transmissions will weaken the wireless signal, and therefore undermine the adversary and enhance security. Moreover, dynamic secrets are lightweight and very suitable to be implemented in resource-limited devices.

In particular, the opportunistic transmission schemes [29, 66], which are studied as energy-efficient wireless communication schemes, naturally promote dynamic secrets based communication security. In these schemes, users only communicate when they find the wireless channel between them to be less random. Suppose the adversary's eavesdropping channel is independent of users' communication channel. When users take advantage of their channel capacity and transmit information at a high bit rate, the adversary may experience a channel fade, i.e. the channel is in a status with high information loss rate, and have many errors in his eavesdropping. The adversary's information loss is then used by dynamic secrets to protect users' communication security.

By assuming that the adversary may have an eavesdropping channel that is worse than users' communication channel from time to time, Refs. [14] and [34] studied the secrecy capacity, i.e. the highest achievable rate of producing perfectly secret, shared key bits for users. Their studies are similar to what we proposed above at the idea level. The difference is, dynamic secrets do not aim to generate perfect secrets. Instead, dynamic secrets are many not-so-perfect secrets. The XORed results of dynamic secrets approach a perfect secret when time goes to infinity. The opportunistic transmission schemes can benefit both dynamic secrets based communication security and the secrecy capacity studied in Refs. [14] and [34].

The combination of energy efficient wireless communications and dynamic secrets based communication security can be used in devices and networks that have energy consumption constraints and demand for communication security, such as wireless body area networks (WBAN) [73] and wireless microsensor networks [48].

Appendix A
Universal-2 Hash Functions

Universal hash refers to a technique that randomly choose a hash function from a class of hash functions, so that the hash output can avoid hash collision as much as possible for an arbitrary or even adversarially constructed input distribution. More precisely, for any fixed input x, if hash function $h(\cdot)$ is randomly chosen from a universal hash function class $\mathcal{H}$, then $h(x)$ is a uniformly distributed random variable.

Universal-2 hash is sometimes called 2-independent universal hash or pairwise independent universal hash. It means for any distinct but fixed x_1 and x_2, $h(x_1)$ and $h(x_2)$ are independent random variables. The mathematic definition of universal-2 hash function is stated as the follows.

Let $f : \mathcal{A} \rightarrow \mathcal{B}$ be a hash function that maps a larger set $\mathcal{A}$ into a smaller set $\mathcal{B}$. x and y are elements in $\mathcal{A}$. $\delta_f(\cdot, \cdot)$ is a collision indicator function.

$$\delta_f(x, y) = \begin{cases} 1 & x \neq y, \ f(x) = f(y) \\ 0 & \textit{otherwise} \end{cases} \tag{A.1}$$

Let $\mathcal{F}$ be a class of hash functions from $\mathcal{A}$ to $\mathcal{B}$. $\delta_{\mathcal{F}}(\cdot, \cdot)$ is defined as

$$\delta_{\mathcal{F}}(x, y) = \sum_{f \in \mathcal{F}} \delta_f(x, y). \tag{A.2}$$

$\mathcal{F}$ is a *universal-2* class of hash functions if for all $x, y \in \mathcal{A}$,

$$\delta_{\mathcal{F}}(x, y) \leq \frac{|\mathcal{F}|}{|\mathcal{B}|}. \tag{A.3}$$

The definition of universal-2 hash functions states that for any distinctive hash inputs x and y, no more than $1/|\mathcal{B}|$ of hash functions in $\mathcal{F}$ will map them into the same hash output.

Reference [25] presented an example of universal-2 function class. Let $|\mathcal{A}| < p$ and $\mathcal{A} = \{0, \ldots, |\mathcal{A}| - 1\}$. p is a prime number. A parameterized hash function f is defined as

$$f_{m,n}(x) \equiv ((mx + n) \bmod p) \bmod |\mathcal{B}|, \tag{A.4}$$

with m, n are integers and $m \neq 0$.

Let $\mathcal{F}$ be a class of hash functions defined by

$$\mathcal{F} = \{f_{m,n} | m \in \{1, \ldots, p-1\}, n \in \{0, \ldots, p-1\}\}. \tag{A.5}$$

$\mathcal{F}$ is a universal-2 hash function class.

We may further note that the hash function class $\mathcal{F}^* = \{f_{m,n} | m \in \{1, \ldots, p-1\}, n = 0\}$ is a universal function class but does not satisfy the criterion of pairwise independent.

Appendix B
Shannon Entropy and Rényi Entropy

Shannon entroy is well known as a measure to the level of randomness contained in a random variable. For a discrete random variable X that is associated with a probability distribution $p(x)$, Its Shannon entropy is

$$H(X) = - \sum_{x \in \{X\}} p(x) \log_2 p(x). \tag{B.1}$$

In the context of information theory, the level of randomness can be interpreted as level of uncertainty when treating X as an information source. Therefore, $H(X)$ also measures the uncertainty of X.

For two random variables X and Y, the conditional Shannon entropy $H(Y|X)$ is defined as

$$H(Y|X) = - \sum_{x \in \{X\}} \sum_{y \in \{Y\}} p_{X,Y}(x, y) \log_2 p_{Y|X}(y|x). \tag{B.2}$$

$H(Y)$ measures the level of uncertainty of information source Y. $H(Y|X)$ measures how much uncertainty left in Y when the information source X is given. If Y is completely determined by X, Y is fixed when X is known. Therefore in this case, $H(Y|X) = 0$. If X carries no information of Y, knowing X does not affect the uncertainty of Y. In this case, $H(Y|X) = H(Y)$. For arbitrary X and Y, we always have $H(Y|X) \leq H(Y)$.

In this monograph, conditional Shannon entropy is used to measure the secrecy level of information. Let Y be a random variable that represents a piece of information, such as a cryptographic key. X be the information gathered by adversary by all means such as through cryptanalysis or eavesdropping. $H(Y|X)$ measures the level of uncertainty that remains in Y from the adversary's view point. In another word, $H(Y|X)$ measures how much secrecy Y has against the adversary.

Rényi entropy is a generalized entropy measure. Rényi entropy of order n is defined for $n \geq 0$ and $n \neq 1$.

S. Xiao et al., *Dynamic Secrets in Communication Security*,
DOI: 10.1007/978-1-4614-7831-7,
© Springer Science+Business Media New York 2014

$$H_n(X) = \frac{1}{1-n} \log_2 \left(\sum_{x \in \{X\}} (p(X = x))^n \right)$$ (B.3)

When $n \rightarrow 1$, Rényi entropy of order n converges to Shannon entropy. In this dissertation, we are particularly interested in Rényi entropy of order 2, which is

$$H_2(X) = -\log_2 \sum_{x \in \{X\}} (p(X = x))^2.$$ (B.4)

Appendix C
Proofs to Theorems in Section 6.5

Theorem 6.2 *Suppose the bit error probability of the binary symmetric channel is* p_e. *the receiver's uncertainty of* $\mathbf{t}$ *given* $\mathbf{r}$ *is lower bounded by*

$$H(\mathbf{t}|\mathbf{r}) \geq \mathbf{n}\mathbf{h}(\mathbf{p_e}) - \mathbf{n}(1 - \mathbf{R})\mathbf{h}(\phi_{nR+1}(\mathbf{p_e})), \tag{C.1}$$

with

$$h(x) = -x \log_2 x - (1 - x) \log_2 (1 - x) \tag{C.2}$$

$$\phi_k(x) = \frac{1 + (1 - 2x)^k}{2}. \tag{C.3}$$

Proof Firstly, we decompose Eq. (C.1).

$$
\begin{aligned}
H(\mathbf{t}|\mathbf{r}) &= H(\mathbf{t}, \mathbf{r}) - H(\mathbf{r}) \\
&= H(\mathbf{t}, \mathbf{t} \oplus \mathbf{r}) - H(\mathbf{r}) \\
&= nR + nh(p_e) - H(\mathbf{r})
\end{aligned} \tag{C.4}
$$

We then divide $\mathbf{r}$ into two parts.

$$\mathbf{r} = \left(\mathbf{r}_s \quad \mathbf{r}_c \right) \tag{C.5}$$

$\mathbf{r}_s$ is the first nR bits and $\mathbf{r}_c$ is the rest $n(1 - R)$ bits. Let $r_c^{(i)}$ represent the ith bits of $\mathbf{r}_c$.

$H(\mathbf{r})$ can be expressed as

$$
\begin{aligned}
H(\mathbf{r}) = H(\mathbf{r}_c, \mathbf{r}_s) &= H(\mathbf{r}_c) + H(\mathbf{r}_c|\mathbf{r}_s) \\
&\leq nR + H(\mathbf{r}_c|\mathbf{r}_s) \\
&\leq nR + \sum_{i=1}^{n(1-R)} H(r_c^{(i)}|\mathbf{r}_s)
\end{aligned} \tag{C.6}
$$

S. Xiao et al., *Dynamic Secrets in Communication Security*,
DOI: 10.1007/978-1-4614-7831-7,
© Springer Science+Business Media New York 2014

The second $\leq$ is because the overlapping of parity check set can only reduce the uncertainty of $\mathbf{r}$.

Let $M^{(i)}$ represent the ith column in the random matrix M of M_G and M_H. $k(i)$ is the number of 1s in $M^{(i)}$.

$$
\begin{aligned}
H(r_c^{(i)}|\mathbf{r}_s) &= E_{\mathbf{r}_s \in \{0,1\}^{nR}}[H(r_c^{(i)}|\mathbf{r}_s)] \\
&= E_{\mathbf{r}_s \in \{0,1\}^{nR}}[h(p(r_c^{(i)} = \mathbf{r}_s M^{(i)}|\mathbf{r}_s))] \\
&= h(\phi_{k(i)+1}(p_e)) \\
&\leq h(\phi_{nR+1}(p_e))
\end{aligned}
\tag{C.7}
$$

The third equality is because M is a equiprobable random binary matrix, $p(r_c^{(i)} = \mathbf{r}_s M^{(i)}|\mathbf{r}_s)$ is the probability of even number of bit errors occur in this parity check set and this probability is unchanged for any $\mathbf{r}_s \in \{0, 1\}^{nR}$.

The last $\leq$ is because $h(\phi_k(p))$ is monotonically increasing with k and $k(i) \leq nR$. Combining the Eqs. (C.4), (C.6) and (C.7) proves the theorem.

Theorem 6.3 *The probability of undetected errors is upper bounded by*

$$
P_{ud} = P_{\mathbf{r} \in \Omega}(\mathbf{t} \neq \mathbf{r}) \leq 2^{-n(1-R)}
\tag{C.8}
$$

Proof Define the error vector $\mathbf{e}$ as

$$
\mathbf{e} = \mathbf{t} \oplus \mathbf{r} = \begin{pmatrix} \mathbf{e}_s & \mathbf{e}_c \end{pmatrix}.
\tag{C.9}
$$

$\mathbf{e}_s$ contains the first nR bits of $\mathbf{e}$. $\mathbf{e}_c$ is the rest.

The syndrome of parity check $\mathbf{s}$ is

$$
\mathbf{s} = \mathbf{e}M_H = \mathbf{e}_s M \oplus \mathbf{e}_c I_n(1 - R).
\tag{C.10}
$$

When $\mathbf{s} = 0$, the received vector $\mathbf{r}$ would be believed as no error-free. The event of undetected error occurs when $\mathbf{e} \neq \mathbf{0}$ and

$$
\mathbf{e}_s M = \mathbf{e}_c.
\tag{C.11}
$$

There are three possible types of errors.

1. $\mathbf{e}_s = 0, \mathbf{e}_c \neq 0$
2. $\mathbf{e}_s \neq 0, \mathbf{e}_c = 0$
3. $\mathbf{e}_s \neq 0, \mathbf{e}_c \neq 0$

Type 1 errors will lead to non-zero syndrome and are always detected. Type 2 and 3 errors may escape the error detection. To analyze the probability of undetected errors, we expand Eq. (C.11) into an equation array.

$$\begin{cases} \mathbf{e}_s M^{(1)} & = e_c^{(1)} \\ \quad\vdots \\ \mathbf{e}_s M^{(i)} & = e_c^{(i)} \\ \quad\vdots \\ \mathbf{e}_s M^{(n(1-R))} & = e_c^{(n(1-R))} \end{cases} \tag{C.12}$$

M is an equiprobable random binary matrix. $M^{(i)}$ is a random binary string picked from $\{0, 1\}^{nR}$. Therefore no matter what the values of $\mathbf{e}_s$ and $e_c^{(i)}$ are, $p(\mathbf{e}_s M^{(i)} = e_c^{(i)}) = \frac{1}{2}$. Because each equation in the array (C.12) is independent to the others, the probability of all equations in (C.12) are satisfied would be $2^{-n(1-R)}$. Therefore we have

$$p_{ud} = p_{\mathbf{r} \in \Omega}(\mathbf{t} \neq \mathbf{r}) \leq 2^{-n(1-R)}. \tag{C.13}$$

The theorem is proved.

Appendix D
Reliability Analysis for Dynamic Key Based Two-Factor Authentication

D.1 MTTR and Average Availability

Recall the four state Markov model for dynamic key based two-factor authentication (Fig. D.1).

The mean time to recover is the expectation of time to stay in B state, which is the reciprocal of the exit rate of this state. Therefore, we have

$$MTTR_{two\text{-}key} = \frac{1}{\mu_c + \mu_m}. \tag{D.1}$$

The average availability is the probability that the system is not in B state in stationary phase. In order to find out the stationary distribution of this model, we first obtain the transition matrix $\mathbf{Q}$.

$$\mathbf{Q} = \begin{pmatrix} -(\lambda_m + \lambda_c) & \lambda_m & \lambda_c & 0 \\ \mu_m & -(\mu_m + \lambda_c) & 0 & \lambda_c \\ \mu_c & 0 & -(\mu_c + \lambda_m) & \lambda_m \\ 0 & \mu_c & \mu_m & -(\mu_c + \mu_m) \end{pmatrix} \tag{D.2}$$

The discrete time transition matrix $\mathbf{P}$ can be derived from $\mathbf{Q}$.

$$\mathbf{P} = \begin{pmatrix} 0 & \dfrac{\lambda_m}{\lambda_m + \lambda_c} & \dfrac{\lambda_c}{\lambda_m + \lambda_c} & 0 \\ \dfrac{\mu_m}{\mu_m + \lambda_c} & 0 & 0 & \dfrac{\lambda_c}{\mu_m + \lambda_c} \\ \dfrac{\mu_c}{\mu_c + \lambda_m} & 0 & 0 & \dfrac{\lambda_m}{\mu_c + \lambda_m} \\ 0 & \dfrac{\mu_c}{\mu_c + \mu_m} & \dfrac{\mu_m}{\mu_c + \mu_m} & 0 \end{pmatrix} \tag{D.3}$$

S. Xiao et al., *Dynamic Secrets in Communication Security*,
DOI: 10.1007/978-1-4614-7831-7,
© Springer Science+Business Media New York 2014

Fig. D.1 Four-state Markov model for dynamic key based two-factor authentication scheme. G_2 state means both keys are secret. G_c state means only the dynamic key on computer is secret. G_m state means only the dynamic key on mobile phone is secret. B state means both keys are known by adversary and the communication security is lost

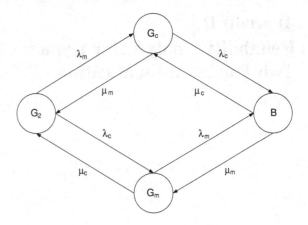

The stationary distribution probability vector $\pi = [\pi_1, \pi_2, \pi_3, \pi_4]$ can be obtained from solving the equations

$$\begin{cases} \pi \mathbf{P} = \pi \\ \pi_1 + \pi_2 + \pi_3 + \pi_4 = 1 \end{cases}.$$ (D.4)

The average availability is

$$A_{avg,two\text{-}key} = 1 - \pi_4 = 1 - \frac{1}{2(1 + \dfrac{\mu_c \mu_m}{\mu_c + \mu_m} \dfrac{\lambda_c + \lambda_m}{\lambda_c \lambda_m})}.$$ (D.5)

D.2 MTFSF

The mean time to first system failure can be calculated from the model shown in Fig. D.2.

In the figure, B state is an absorbing state. We use T_{2B}, T_{cB}, and T_{mB} to denote the average time to reach the absorbing state from state G_2, G_c, and G_m respectively. T_{2B}, T_{cB}, and T_{mB} are related by the following equation array.

$$\begin{cases} T_{cB} = \dfrac{1}{\mu_m + \lambda_c} + \dfrac{\mu_m}{\mu_m + \lambda_c} T_{2B} \\ T_{mB} = \dfrac{1}{\mu_c + \lambda_m} + \dfrac{\mu_c}{\mu_c + \lambda_m} T_{2B} \\ T_{2B} = \dfrac{1}{\lambda_c + \lambda_m} + \dfrac{\lambda_m}{\lambda_c + \lambda_m} T_{cB} + \dfrac{\lambda_c}{\lambda_c + \lambda_m} T_{mB} \end{cases}$$ (D.6)

Fig. D.2 Four-state Markov model to calculate MTFSF for dynamic key based two-factor authentication scheme. G_2 state means both keys are secret. G_c state means only the dynamic key on computer is secret. G_m state means only the dynamic key on mobile phone is secret. B state means both keys are known by adversary and the communication security is lost

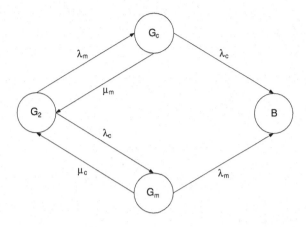

The MTFSF of the authentication scheme can be found by solving the above equation array. We have

$$MTFSF_{two\text{-}key} = T_{2B} = \frac{1 + \dfrac{\lambda_m}{\mu_m + \lambda_c} + \dfrac{\lambda_c}{\mu_c + \lambda_m}}{\lambda_c \lambda_m \left(\dfrac{1}{\mu_m + \lambda_c} + \dfrac{1}{\mu_c + \lambda_m} \right)}. \tag{D.7}$$

References

1. Adams, C., Lloyd, S.: Understanding PKI: Concepts, Standards, and Deployment Considerations, 2nd edn. Addison-Wesley Longman Publishing, Boston (2002)
2. Al-Bakri, S., Kiah, M., Zaidan, A., Zaidan, B., Alam, G.: Securing peer-to-peer mobile communications using public key cryptography: new security strategy. Int. J. Phys. Sci. **6**(4), 930–938 (2011)
3. Al-Janabi, S., Rasheed, M.S.: Public-key cryptography enabled kerberos authentication. In: Developments in E-systems Engineering (DeSE), 2011, pp. 209–214 (2011)
4. Amin, M.: Challenges in reliability, security, efficiency, and resilience of energy infrastructure: toward smart self-healing electric power grid. In: Power and Energy Society General Meeting—Conversion and Delivery of Electrical Energy in the 21st Century, 2008 IEEE, pp. 1–5 (2008)
5. Anderson, R.J. (ed.): Fast Software Encryption, Cambridge Security Workshop, Cambridge, UK, December 9–11, 1993, Proceedings. Lecture Notes in Computer Science, vol. 809. Springer (1994)
6. Andersson, G., Donalek, P., Farmer, R., Hatziargyriou, N., Kamwa, I., Kundur, P., Martins, N., Paserba, J., Pourbeik, P., Sanchez-Gasca, J., Schulz, R., Stankovic, A., Taylor, C., Vittal, V.: Causes of the 2003 major grid blackouts in North America and Europe, and recommended means to improve system dynamic performance. IEEE Trans. Power Syst. **20**(4), 1922–1928 (2005)
7. Andgren, M., Hell, M., Johansson, T.: On hardware-oriented message authentication with applications towards RFID. In: Lightweight Security Privacy: Devices, Protocols and Applications (LightSec), 2011 Workshop on, pp. 26–33 (2011)
8. Aono, T., Higuchi, K., Ohira, T., Komiyama, B., Sasaoka, H.: Wireless secret key generation exploiting reactance-domain scalar response of multipath fading channels. IEEE Trans. Antenn. Prop. **53**(11), 3776–3784 (2005)
9. Arfman, J.M., Roden, P.: Project Athena: supporting distributed computing at MIT. IBM Syst. J. **31**(3), 550–563 (1992)
10. Azimi-Sadjadi, B., Kiayias, A., Mercado, A., Yener, B.: Robust key generation from signal envelopes in wireless networks. In: CCS '07: Proceedings of the 14th ACM Conference on Computer and Communications Security, pp. 401–410. ACM, New York, NY, USA (2007)
11. Bardou, R., Focardi, R., Kawamoto, Y., Simionato, L., Steel, G., Tsay, J.K.: Efficient padding oracle attacks on cryptographic hardware. In: Safavi-Naini, R., Canetti, R. (eds.) Advances in Cryptology, CRYPTO 2012. Lecture Notes in Computer Science, vol. 7417, pp. 608–625. Springer, Berlin (2012)

S. Xiao et al., *Dynamic Secrets in Communication Security*,
DOI: 10.1007/978-1-4614-7831-7,
© Springer Science+Business Media New York 2014

12. Barker, W.C.: Recommendation for the triple data encryption algorithm (TDEA) block cipher. Technical Report, NIST (2008)
13. Barlow, R., Proschan, F., Hunter, L.: Mathematical theory of reliability. Classics in applied mathematics. SIAM (1996)
14. Barros, J., Rodrigues, M.R.D.: Secrecy capacity of wireless channels. Information Theory, 2006 IEEE International Symposium on, pp. 356–360 (2006)
15. Bauer, M., Plappert, W., Wang, C., Dostert, K.: Packet-oriented communication protocols for smart grid services over low-speed PLC. In: Power Line Communications and Its Applications, 2009. ISPLC 2009. IEEE International Symposium on, pp. 89–94 (2009)
16. Baumeister, T.: Adapting PKI for the smart grid. In: Smart Grid Communications (SmartGridComm), 2011 IEEE International Conference on, pp. 249–254 (2011)
17. Bell, J.: Mars exploration: roving the red planet. Nature **490**(7418), 34–35 (2012)
18. Bello, L.: DSA-1571-1 openssl—predictable random number generator. Tech. rep., www. Debian.org (2008)
19. Bennett, C.H., Bessette, F., Brassard, G., Salvail, L., Smolin, J.: Experimental quantum cryptography. J. Cryptol. **5**, 3–28 (1992)
20. Bennett, C.H., Brassard, G.: Quantum cryptography: public key distribution and coin tossing. In: Proceedings of the IEEE International Conference on Computers, Systems and Signal Processing, pp. 175–179. IEEE Press, New York (1984)
21. Bennett, C.H., Brassard, G., Crkpeau, C., Maurer, U.M.: Generalized privacy amplification. IEEE Trans. Inf. Theory **41**, 1915–1923 (1995)
22. Bennett, C.H., Brassard, G., Robert, J.M.: Privacy amplification by public discussion. SIAM J. Comput. **17**(2), 210–229 (1988)
23. Bittau, A., Handley, M., Lackey, J.: The final nail in WEP's coffin. In: Security and Privacy, 2006 IEEE Symposium on, pp. 15–400. doi:10.1109/SP.2006.40 (2006)
24. Bloch, M., Barros, J., Rodrigues, M.R.D., McLaughlin, S.W.: Wireless information-theoretic security, Theoretical aspects. IEEE Trans. Inf. Theory, **54**(6), 2515–2534 (2008)
25. Carter, J.L., Wegman, M.N.: Universal classes of hash functions (extended abstract). In: Proceedings of the Ninth Annual ACM Symposium on Theory of Computing, STOC '77, pp. 106–112. ACM, New York, NY, USA (1977)
26. Carter, J.L., Wegman, M.N.: Universal classes of hash functions. J. Comput. Syst. Sci. **18**, 396–407 (1979)
27. de Carvalho, J., Veiga, H., Gomes, P., Pacheco, C., Marques, N., Reis, A.: Laboratory performance of wi-fi point-to-point links: a case study. In: Wireless Telecommunications Symposium, 2009. WTS 2009, pp. 1–5 (2009)
28. Chevassut, O., Alain Fouque, P., Gaudry, P., Pointcheval, D.: Key derivation and randomness extraction. In: Proceedings of Crypto'05 (2005)
29. Chong, Z., Jorswieck, E.: Energy efficiency in random opportunistic beamforming. In: Vehicular Technology Conference (VTC Spring), 2011 IEEE 73rd, pp. 1–5 (2011)
30. Cover, T.M., Thomas, J.A.: Elements of Information Theory. Wiley-Interscience, New York (1991)
31. Cronin, E., Sherr, M., Blaze, M.: On the (un)reliability of eavesdropping. Int. J. Secur. Netw. (IJSN) **3**(2), 103–113 (2008)
32. Deconinck, G.: An evaluation of two-way communication means for advanced metering in flanders (Belgium). In: Instrumentation and Measurement Technology Conference Proceedings, 2008. IMTC 2008. IEEE, pp. 900–905 (2008)
33. Delsing, J., Eliasson, J., Leijon, V.: Latency and packet loss of an interferred 802.15.4 channel in an industrial environment. In: Sensor Technologies and Applications (SENSORCOMM), 2010 Fourth International Conference on, pp. 33–38 (2010)
34. Ding, Z., Leung, K., Goeckel, D., Towsley, D.: On the application of cooperative transmission to secrecy communications. IEEE J. Sel. Areas Commun. **30**(2), 359–368 (2012)
35. Dorrendorf, L., Gutterman, Z., Pinkas, B.: Cryptanalysis of the random number generator of the windows operating system. ACM Trans. Inf. Syst. Secur. **13**(1), 10:1–10:32 (2009)

36. Dutta, P., Dawson-Haggerty, S., Chen, Y., Liang, C.J.M., Terzis, A.: Design and evaluation of a versatile and efficient receiver-initiated link layer for low-power wireless. In: Proceedings of the 8th ACM Conference on Embedded Networked Sensor Systems, SenSys '10, pp. 1–14. ACM, New York, NY, USA (2010)

37. Finke, T., Gebhardt, M., Schindler, W.: A new side-channel attack on RSA prime generation. In: Clavier, C., Gaj, K. (eds.) Cryptographic Hardware and Embedded Systems—CHES 2009. Lecture Notes in Computer Science, vol. 5747, pp. 141–155. Springer, Berlin (2009)

38. Flatraaker, D.I.: Mobile, internet and electronic payments: the key to unlocking the full potential of the internal payments market. J. Paym. Strategy Syst. 3(1), 60–70 (2009)

39. Florencio, D., Herley, C.: A large-scale study of web password habits. In: WWW '07: Proceedings of the 16th international Conference on World Wide Web, pp. 657–666. ACM, New York, NY, USA (2007)

40. Floyd, S., Fall, K.: Promoting the use of end-to-end congestion control in the internet. IEEE/ACM Trans. Netw. 7(4), 458–472 (1999)

41. Frst, M., Weier, H., Nauerth, S., Marangon, D.G., Kurtsiefer, C., Weinfurter, H.: High speed optical quantum random number generation. Optics Express 18, 13,029–13,037 (2010)

42. Furnell, S., Clarke, N., Karatzouni, S.: Beyond the pin: enhancing user authentication for mobile devices. Comput. Fraud Secur. 2008(8), 12–17 (2008)

43. Gafurov, D., Snekkenes, E., Buvarp, T.: Robustness of biometric gait authentication against impersonation attack. In: Meersman, R., Tari, Z., Herrero, P. (eds.) On the Move to Meaningful Internet Systems 2006: OTM 2006 Workshops. Lecture Notes in Computer Science, vol. 4277, pp. 479–488. Springer, Berlin (2006)

44. Goeckel, D., Vasudevan, S., Towsley, D., Adams, S., Ding, Z., Leung, K.: Artificial noise generation from cooperative relays for everlasting secrecy in two-hop wireless networks. IEEE J. Sel. Areas Commun. 29(10), 2067–2076 (2011)

45. Gollmann, D. (ed.): Fast Software Encryption, Third International Workshop, Cambridge, UK, 21–23 February 1996, Proceedings. Lecture Notes in Computer Science, vol. 1039. Springer, Berlin (1996)

46. Guneysu, T., Kasper, T., Novotny, M., Paar, C., Rupp, A.: Cryptanalysis with COPACOBANA. IEEE Trans. Comput. 57(11), 1498–1513 (2008)

47. Hassan, A.A., Stark, W.E., Hershey, J.E.: Cryptographic key agreement for mobile radio. Dig. Sig. Process. 6, 207–212 (1996)

48. Heinzelman, W. R., Chandrakasan, A., Balakrishnan, H.: Energy-efficient communication protocol for wireless microsensor networks. In: System Sciences, 2000. Proceedings of the 33rd Annual Hawaii International Conference on, vol. 2, p. 10 (2000)

49. Herley, C.: So long, and no thanks for the externalities: the rational rejection of security advice by users. In: NSPW '09: Proceedings of the 2009 Workshop on New Security Paradigms Workshop, pp. 133–144. ACM, New York, NY, USA (2009)

50. Honan, M.: How apple and amazon security flaws led to my epic hacking

51. IEEE: IEEE standard for information technology-telecommunications and information exchange between systems-local and metropolitan area networks-specific requirements—Part 11: Wireless LAN medium access control (MAC) and physical layer (phy) specifications. IEEE Std 802.11-2007 (Revision of IEEE Std 802.11-1999) pp. C1–1184 (2007)

52. Jagatic, T.N., Johnson, N.A., Jakobsson, M., Menczer, F.: Social phishing. Commun. ACM 50(10), 94–100 (2007)

53. Jana, S., Premnath, S.N., Clark, M., Kasera, S.K., Patwari, N., Krishnamurthy, S.V.: On the effectiveness of secret key extraction from wireless signal strength in real environments. In: Proceedings of the 15th Annual International Conference on Mobile Computing and Networking, MobiCom '09, pp. 321–332. ACM, New York, NY, USA (2009)

54. Janssen, J., De Vleeschauwer, D., Buchli, M., Petit, G.: Assessing voice quality in packet-based telephony. IEEE Internet Comput. 6(3), 48–56 (2002)

55. Jonsson, E., Olovsson, T.: An empirical model of the security intrusion process. In: Computer Assurance, 1996. COMPASS '96, 'Systems Integrity, Software Safety, and Process Security'. Proceedings of the Eleventh Annual Conference on, pp. 176–186 (1996)

56. Jonsson, E., Olovsson, T.: A quantitative model of the security intrusion process based on attacker behavior. IEEE Trans. Softw. Eng. **23**(4), 235–245 (1997)
57. Kentros, S., Albayram, Y., Bamis, A.: Towards macroscopic human behavior based authentication for mobile transactions. In: Proceedings of the 2012 ACM Conference on Ubiquitous Computing, pp. 641–642. ACM (2012)
58. Kerckhoffs, A.: La cryptographie militaire. J. Sci. Militaires **IX**, 5–83 (1883)
59. Key, P., Massoulié, L., Towsley, D.: Path selection and multipath congestion control. Commun. ACM **54**(1), 109–116 (2011)
60. Han H, Srinivas S, C. V. Hollot, R. Srikant, and Donald Towsley. Multi-path TCP: a joint congestion control and routing scheme to exploit path diversity in the internet. IEEE/ACM Trans. Netw. 14(6), 1260–1271 (2006)
61. Khurana, H., Hadley, M., Lu, N., Frincke, D.: Smart-grid security issues. IEEE Secur. Priv. **8**(1), 81–85 (2010)
62. Kinney, R., Crucitti, P., Albert, R., Latora, V.: Modeling cascading failures in the North American power grid. Eur. Phys. J. B: Condens. Matter Complex Syst. **46**, 101–107 (2005)
63. Klein, A.: Attacks on the RC4 stream cipher. Des. Codes Crypt. **48**, 269–286 (2008)
64. Koch, R., Stelte, B., Golling, M.: Attack trends in present computer networks. In: Cyber Conflict (CYCON), 2012 4th International Conference on, pp. 1–12 (2012)
65. Korhonen, J., Wang, Y.: Effect of packet size on loss rate and delay in wireless links. In: Wireless Communications and Networking Conference, 2005 IEEE, vol. 3, pp. 1608–1613 (2005)
66. Kouyoumdjieva, S., Helgason, I., Yavuz, E.A., Karlsson, G.: Evaluating an energy-efficient radio architecture for opportunistic communication. In: Proceedings of 3rd Workshop on Energy Efficiency in Wireless Networks and Wireless Networks for Energy Efficiency (E2Nets) (2012). QC 20120803
67. Krawczyk, H.: LFSR-based hashing and authentication. In: Desmedt, Y. (ed.) Advances in Cryptology. CRYPTO 94. Lecture Notes in Computer Science, vol. 839, pp. 129–139. Springer, Berlin (1994)
68. Kurita, S., Komoriya, K., Uda, R.: Privacy protection on transfer system of automated teller machine from brute force attack. In: Advanced Information Networking and Applications Workshops (WAINA), 2012 26th International Conference on, pp. 72–77 (2012)
69. Laptyeva, T.V., Flach, S., Kladko, K.: The weak password problem: chaos, criticality, and encrypted p-CAPTCHAs. Cryptology ePrint Archive, Report 2011/172 (2011)
70. Latif, M.A., Sultan, A., Gamal, H.E.: ARQ-based secret key sharing. In: ICC, pp. 1–6 (2009)
71. Latif, M.A., Sultan, A., Gamal, H.E.: ARQ secrecy over correlated fading channels. In: Information Theory Workshop, 2010. ITW 2010. IEEE (2010)
72. Lefebvre, S., Porteous, H.: The Russian 10.. 11: an inconsequential adventure? Int. J. Intell. Counter Intell. **24**(3), 447–466 (2011)
73. Li, C., Li, H., Kohno, R.: Performance evaluation of IEEE 802.15. 4 for wireless body area network (WBAN). In: Communications Workshops, 2009. ICC Workshops 2009. IEEE International Conference on, pp. 1–5. IEEE (2009)
74. Li, H., Han, Z., Lai, L., Qiu, R., Yang, D.: Efficient and reliable multiple access for advanced metering in future smart grid. In: Smart Grid Communications (SmartGridComm), 2011 IEEE International Conference on, pp. 440–444 (2011)
75. Li, Y., Xiong, Y., Yang, S.: Study on mobile commerce authentication system. In: Wireless Communications, Networking and Mobile Computing (WiCOM), 2011 7th International Conference on, pp. 1–4 (2011)
76. Liu, R., Liu, T., Poor, H.V., Shamai, S.: Multiple-input multiple-output Gaussian broadcast channels with confidential messages. CoRR **abs/0903.3786** (2009)
77. Liu, R., Poor, H.: Multi-antenna Gaussian broadcast channels with confidential messages. In: ISIT 2008. IEEE International Symposium on Information Theory, 2008 (2008)
78. Lobato, R., Thomas, J.: The business of anti-piracy: new zones of enterprise in the copyright wars. Int. J. Commun. **5** (2011)

79. Long, H., Liu, Y., Fan, X., Dick, R.P., Yang, H.: Energy-efficient spatially-adaptive clustering and routing in wireless sensor networks. In: Proceedings of the Conference on Design, Automation and Test in Europe, DATE '09, pp. 1267–1272. European Design and Automation Association, 3001 Leuven, Belgium, Belgium (2009)

80. Lumezanu, C., Guo, K., Spring, N., Bhattacharjee, B.: The effect of packet loss on redundancy elimination in cellular wireless networks. In: Proceedings of the 10th Annual Conference on Internet Measurement, IMC '10, pp. 294–300. ACM, New York, NY, USA (2010)

81. Main, A., van Oorschot, P.: Software protection and application security: understanding the battleground. International Course on State of the Art and Evolution of Computer Security and Industrial Cryptography, Heverlee, Belgium (2003)

82. Mathur, S., Trappe, W., Mandayam, N., Ye, C., Reznik, A.: Radio-telepathy: extracting a secret key from an unauthenticated wireless channel. In: MobiCom '08: Proceedings of the 14th ACM International Conference on Mobile Computing and Networking, pp. 128–139. ACM, New York, NY, USA (2008)

83. Maurer, U.M.: Secret key agreement by public discussion from common information. IEEE Trans. Inf. Theory **39**, 733–742 (1993)

84. Maurer, U.M., Wolf, S.: Secret key agreement over a non-authenticated channel—Part I: Definitions and bounds. IEEE Trans. Inf. Theory **49**, 822–831 (2003)

85. Maurer, U.M., Wolf, S.: Secret key agreement over a non-authenticated channel—Part II: The simulatability condition. IEEE Trans. Inf. Theory **49**, 832–838 (2003)

86. Maurer, U.M., Wolf, S.: Secret key agreement over a non-authenticated channel—Part III: Privacy amplification. IEEE Trans. Inf. Theory **49**, 839–851 (2003)

87. Melia-Segui, J., Garcia-Alfaro, J., Herrera-Joancomarti, J.: Analysis and improvement of a pseudorandom number generator for EPC Gen2 tags. In: Sion, R., Curtmola, R., Dietrich, S., Kiayias, A., Miret, J., Sako, K., Seb, F. (eds.) Financial Cryptography and Data Security. Lecture Notes in Computer Science, vol. 6054, pp. 34–46. Springer, Berlin (2010)

88. Miao, G., Himayat, N., Li, Y.G., Swami, A.: Cross-layer optimization for energy-efficient wireless communications: a survey. Wireless Commun. Mobile Comput. **9**(4), 529–542 (2009)

89. Miller, F., Vandome, A., McBrewster, J.: Key-Agreement Protocol. VDM Verlag Dr. Mueller e.K. (2010)

90. Naumovich, G., Memon, N.: Preventing piracy, reverse engineering, and tampering. Computer **36**(7), 64–71 (2003)

91. NIST: Federal information processing standards publication 197. Technical Report, NIST (2001)

92. Omar, Y., Youssef, M., El Gamal, H.: ARQ secrecy: from theory to practice. In: Information Theory Workshop, 2009. ITW 2009. IEEE, pp. 6–10 (2009)

93. Pareschi, F., Scotti, G., Giancane, L., Rovatti, R., Setti, G., Trifiletti, A.: Power analysis of a chaos-based random number generator for cryptographic security. In: Circuits and Systems, 2009. ISCAS 2009. IEEE International Symposium on, pp. 2858–2861 (2009)

94. Paxson, V.: End-to-end routing behavior in the internet. SIGCOMM Comput. Commun. Rev. **26**(4), 25–38 (1996)

95. Piètre-Cambacédès, L., Sitbon, P.: Cryptographic key management for SCADA systems-issues and perspectives. In: Proceedings of the 2008 International Conference on Information Security and Assurance (ISA 2008), ISA '08, pp. 156–161. IEEE Computer Society, Washington, DC, USA (2008)

96. Piller, C.: How piracy opens doors for windows. Los Angeles Times **9** (2006)

97. Podhoransky, P., Lipovsky, M., Zemanovic, J., Sabo, M.: Transfer and error rate measurement in the Lon works power line communication systems. In: Radioelektronika, 2007. 17th International Conference, pp. 1–3 (2007)

98. Power, R., Forte, D.: Social engineering: attacks have evolved, but countermeasures have not. Comput. Fraud Secur. **2006**(10), 17–20 (2006)

99. Raghuvamshi, A., Rao, P.: An effortless cryptanalytic attack on knapsack cipher. In: Process Automation, Control and Computing (PACC), 2011 International Conference on, pp. 1–6 (2011)

100. Rana, M., Ahmed, K., Sumel, N., Alam, M., Sarkar, L.: Security in ad hoc networks: a location based impersonation detection method. In: Computer Engineering and Technology, 2009. ICCET '09. International Conference on, vol. 2, pp. 380–384 (2009)

101. Rappaport, T.: Wireless Communications: Principles and Practice. Prentice Hall PTR, Upper Saddle River (2001)

102. Ren, Y., Xing, T., Cao, G., Xu, E., Chen, X.: Research and practice on the cooperative concealing technology of trojan horses. In: Networking and Digital Society (ICNDS), 2010 2nd International Conference on, vol. 1, pp. 216–219 (2010)

103. Renyi, A.: On measures of information and entropy. In: Proceedings of the 4th Berkeley Symposium on Mathematics, Statistics and Probability (1960)

104. Rivest, R.L.: The RC4 Encryption Algorithm. RSA Data Security, Inc. (1992)

105. Romero-Jerez, J., Goldsmith, A.: Receive antenna array strategies in fading and interference: an outage probability comparison. IEEE Trans. Wireless Commun. **7**(3), 920–932 (2008)

106. Romirer-Maierhofer, P., Ricciato, F., D'Alconzo, A., Franzan, R., Karner, W.: Network-wide measurements of TCP RTT in 3G. In: TMA, pp. 17–25 (2009)

107. Rozema, L., Darabi, A., Mahler, D., Hayat, A., Soudagar, Y., Steinberg, A.M.: Direct violation of Heisenberg's precision limit by weak measurements. In: Frontiers in Optics Conference, p. FW4J.4. Optical Society of America (2012)

108. Salem, M., Stolfo, S.: Modeling user search behavior for masquerade detection. In: Sommer, R., Balzarotti, D., Maier, G. (eds.) Recent Advances in Intrusion Detection. Lecture Notes in Computer Science, vol. 6961, pp. 181–200. Springer, Berlin (2011)

109. Sathish Babu, B., Venkataram, P.: A dynamic authentication scheme for mobile transactions. Int. J. Netw. Secur. **8**(1), 59–74 (2009)

110. Schneier, B.: Applied Cryptography: Protocols, Algorithms, and Source Code in C. Wiley, New York (1995)

111. Serrano, P., Zink, M., Kurose, J.: Assessing the fidelity of cots 802.11 sniffers. In: INFOCOM 2009, IEEE, pp. 1089–1097 (2009)

112. Shafiq, M.Z., Ji, L., Liu, A.X., Pang, J., Wang, J.: A first look at cellular machine-to-machine traffic: large scale measurement and characterization. SIGMETRICS Perform. Eval. Rev. **40**(1), 65–76 (2012)

113. Shannon, C.E.: Communication theory of secrecy systems. Bell Syst. Tech. J. **28**, 656–715 (1949)

114. Sheth, A., Nedevschi, S., Patra, R., Surana, S., Brewer, E., Subramanian, L.: Packet loss characterization in wifi-based long distance networks. In: INFOCOM 2007. 26th IEEE International Conference on Computer Communications. IEEE, pp. 312–320 (2007)

115. Siam, M., Krunz, M., Cui, S., Muqattash, A.: Energy-efficient protocols for wireless networks with adaptive MIMO capabilities. Wireless Netw. **16**, 199–212 (2010)

116. Sieka, B.: Using radio device fingerprinting for the detection of impersonation and Sybil attacks in wireless networks. In: Buttyn, L., Gligor, V., Westhoff, D. (eds.) Security and Privacy in Ad-Hoc and Sensor Networks. Lecture Notes in Computer Science, vol. 4357, pp. 179–192. Springer, Berlin (2006)

117. Singh, S.: The Code Book: The Evolution of Secrecy from Mary, Queen of Scots, to Quantum Cryptography, 1st edn. Doubleday, New York (1999)

118. Smith, S.: Cryptographic scalability challenges in the smart grid (extended abstract). In: Innovative Smart Grid Technologies (ISGT), 2012 IEEE PES, pp. 1–3 (2012)

119. Stevens, R.W.: Unix Network Programming. Prentice Hall PTR, Upper Saddle River (1990)

120. Tanenbaum, A.: Computer networks. Prentice Hall PTR, Upper Saddle River (2003)

121. Tang, X., Liu, R., Spasojevic, P., Poor, H.: On the throughput of secure hybrid-ARQ protocols for Gaussian block-fading channels. IEEE Trans. Inf. Theory **55**(4), 1575–1591 (2009)

122. Tao, Z., Nath, B., Lonie, A.: A data clustering approach to discriminating impersonating devices in wi-fi networks. Secur. Commun. Netw. **3**(1), 44–57 (2010)

123. Thornburgh, T.: Social engineering: the "dark art". In: Proceedings of the 1st Annual Conference on Information Security Curriculum Development, InfoSecCD '04, pp. 133–135. ACM, New York, NY, USA (2004)

124. Viega, J.: Practical random number generation in software. In: Computer Security Applications Conference, 2003. Proceedings. 19th Annual, pp. 129–140 (2003)
125. Vishnani, K., Pais, A.R., Mohandas, R.: An in-depth analysis of the epitome of online stealth: keyloggers; and their countermeasures. In: Abraham, A., Mauri, J.L., Buford, J.F., Suzuki, J., Thampi, S.M. (eds.) Advances in Computing and Communications. Communications in Computer and Information Science, vol. 192, pp. 10–19. Springer, Berlin (2011)
126. Weir, M., Aggarwal, S., Collins, M., Stern, H.: Testing metrics for password creation policies by attacking large sets of revealed passwords. In: Proceedings of the 17th ACM Conference on Computer and Communications Security, CCS '10, pp. 162–175. ACM, New York, NY, USA (2010)
127. Wilson, R., Tse, D., Scholtz, R.: Channel identification: secret sharing using reciprocity in ultrawideband channels. In: ICUWB 2007, pp. 270–275 (2007)
128. Wong, C.W., Shea, J., Wong, T.: Secret sharing in fast fading channels based on reliability-based hybrid ARQ. In: MILCOM 2008. IEEE, pp. 1–7 (2008)
129. Xiao, S., Pishro-Nik, H., Gong, W.: Dense parity check based secrecy sharing in wireless communications. In: Global Telecommunications Conference, 2007. GLOBECOM '07. IEEE, pp. 54–58 (2007)
130. Yang, C., Hung, J.L., Lin, Z.X.: Loose password security in chinese cyber world left the front door wide open to hackers: an analytic view. In: Proceedings of the 14th Annual International Conference on Electronic Commerce, ICEC '12, pp. 121–126. ACM, New York, NY, USA (2012)
131. Yang, D., Sonmez, M., Bosworth, D., Fryxell, G.: Global software piracy: searching for further explanations. J. Bus. Ethics **87**(2), 269–283 (2009)
132. Yin, J., Ren, J.G., Lu, H., Cao, Y., Yong, H.L., Wu, Y.P., Liu, C., Liao, S.K., Zhou, F., Jiang, Y.: Quantum teleportation and entanglement distribution over 100-kilometre free-space channels. Nature **488**(7410), 185188 (2012). http://www.nature.com/nature/journal/v488/n7410/abs/nature11332.html
133. Yurcik, W., Liu, C.: A first step toward detecting SSH identity theft in HPC cluster environments: discriminating masqueraders based on command behavior. In: Cluster Computing and the Grid, 2005. CCGrid 2005. IEEE International Symposium on, vol. 1, pp. 111–120 (2005)
134. Zeng, K., Govindan, K., Wu, D., Mohapatra, P.: Identity-based attack detection in mobile wireless networks. In: INFOCOM, 2011 Proceedings IEEE, pp. 1880–1888 (2011)
135. Zhao, S., Shoniregun, C.: Critical review of unsecured WEP. In: Services, 2007 IEEE Congress on, pp. 368–374 (2007)
136. Zhu, W., Thomborson, C., Wang, F.: A survey of software watermarking. Intelligence and Security Informatics, pp. 283–331 (2005)

Index

S. Xiao et al., *Dynamic Secrets in Communication Security*,
DOI: 10.1007/978-1-4614-7831-7,
© Springer Science+Business Media New York 2014

135

Printed in the United States
By Bookmasters